HERBARIUM

HERBARIUM

Manel Armengol

Text by Margaret Hooks

TURNER

Manel Armengol:
Seeing the world in a grain of sand

When Manel Armengol began work on the "Herbarium" series in 2004, he did so with the intention of once again depicting the essential components of the world we inhabit, this time within the botanical realm of a simple meadow in a corner of his native Catalunya. In creating the images reproduced in this book he utilized the camera to pare down his subjects to their essence, evoking an ecstatic aesthetic rooted in austerity, a leitmotif which is at the core of a body of work spanning two decades.

Armengol's trajectory intricately weaves life and art making it impossible to write about one without referencing the other. He first began taking photographs in 1975 to accompany the articles he wrote as a news reporter. Within a year however, his medium was no longer words, but had become images; images that soon earned him a reputation as an intrepid press photographer when, following the death of Spanish dictator General Francisco Franco, he captured the spirit of Spain at a critical point in its history making hundreds of photographs of women, workers and students who took to the streets to flex the muscle of a tenuous new-found freedom.

His images of this peaceful rebellion, and the authorities' violent response, traveled around the world on the pages of *Paris-Match*, the *New York Times*, *Newsweek* and *Der Spiegel* and came to symbolize Spain's transition from fascism to democracy. Over the next five years Armengol traversed the globe in pursuit of images that established him as an exceptional, prolific photojournalist. In 1981, a hiatus in his work occurred after a catastrophic car accident that marked a decisive turning point in his life. Returning late at night from dropping off some photographs at a publication he was working for, the car he was driving in was broadsided by another vehicle. Seriously injured, he hovered between life and death for several mind bending moments before being transported to the hospital where he was interned for several weeks.

Upon leaving the hospital, he went home to an apartment he had just leased in one of the world's outstanding architectural structures, the inimitable Casa Mila, a creation of the visionary Catalan architect, Antonio Gaudi. Faced with a lengthy convalescence, he was at least fortunate that he had found not only the perfect place in which to recuperate, but one that would positively, and irrevocably affect his life and work. His injuries soon began to heal, however, the "near-death" experience that occurred during his accident continued to have an intense effect on his psyche. Fearing that it could eventually overpower him, he struggled to come to terms with it. Drowning out the everyday world with Gregorian music, he immersed himself in the written word, avidly devouring volume after esoteric volume of works on mythology, religion and mysticism.

Submerged in the unique space of his apartment in the Casa Mila, his imagination expanded to fill the white, womblike space, with its rounded, organic forms and surreal, three dimensional ceilings covered in patterns of diverse shapes and inscriptions. Moulded in plaster on the ceiling of

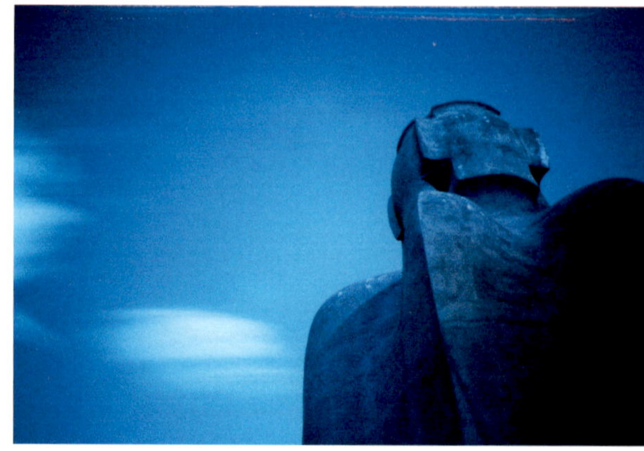

Trust. "Gaudi at Night" series
Barcelona, 1983
Cibachrome 14,2 x 21,6 in

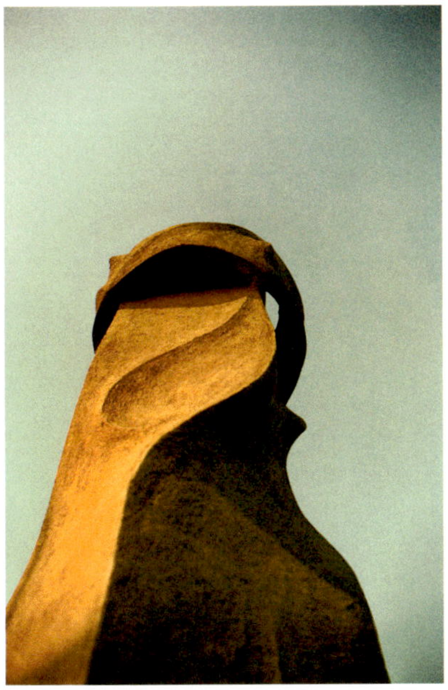

his bedroom was inscribed just one awe-inspiring word –"Terrible"– a word that resonated throughout the space and spoke to him of profound change; of extreme difficulty and transformation in his life.

Built between 1906-1912 the Casa Mila was the last civic building Gaudi built. Many now consider it to be his best; certainly it was his most controversial. A triumph of great skill and imagination, the building was not conceived as a simple dwelling, but rather a work of major structural innovations. These coupled with the architect's development of a ground-breaking, wholly organic concept of architectural creation, made it the first building to transcend architecture and take the form of an abstract sculpture, or perhaps more aptly – an early precursor of the three-dimensional, surrealist work of art.

An accolade to the curvilinear, progeny of a passionate affair with the rotund, the Casa Mila's rippling, undulating façade wraps and rolls around an entire city block. And Gaudi had literally crowned his achievement with his most spectacular feat; a roof terrace of fabulous complexity with chimney pots, skylights and ventilators a pretext for the strange creatures and forms of his fantasy.

It was on this rooftop almost three years into his convalescence, with no income and few prospects of work that Armengol had his "dark night of the soul" that resulted in an epiphany. He often ascended to the roof terrace at dusk to contemplate Gaudi's disconcerting, grotesque landscape populated with life-like forms most of them macabre: warriors of past and future wars; armoured medieval knights, blank-faced killing automatons and perhaps most chilling; a ghastly, looming figure its head a gaping, petrified mask.

Oracle. "Gaudi at Night" series
Barcelona, 1983
Cibachrome 21.6 x 14.2 in

This monstrous assembly mirrored the demons he did battle with daily as he struggled to emerge intact from the terrifying experience of the accident. One evening as he climbed up to the roof and surveyed the extraordinary panorama it occurred to him to do the one thing he knew how to do best – he would photograph it. He rushed downstairs to find camera and film, but the only film he could locate was at the back of his refrigerator and most of it was long past the expiration date. Undaunted he loaded it into the camera and went back up to the rooftop, set up his tripod, and waited, and went on waiting, for as long as was necessary in order to make the lengthy nocturnal exposures he wanted.

He did this night after night, often working until dawn, and when he was finished the result was an outstanding body of work that transcends the usual depictions of these Gaudi sculptural forms. In his photographs the fearful geometry of the figures is transformed in the alchemy of an artist's vision. Bathed in the eerie light of his time-stretching exposures, and depicted in a manner the human eye is incapable of perceiving, the bizarre forms appear to inhabit another dimension, another phantasmagorical world.

As he waited through the long, dark hours to obtain the images he desired Armengol went through a powerful transformation that altered his perspective on life and how he thought of himself as a photographer. As he described it, he no longer considered himself a "hunter" on the prowl for a good shot and he knew he would never again be an action photographer. Not because his injuries meant it had become more difficult for him to get around as quickly as needed, but because on another, deeper plane his was now a measured, more unhurried pace, in both his life, and his art.

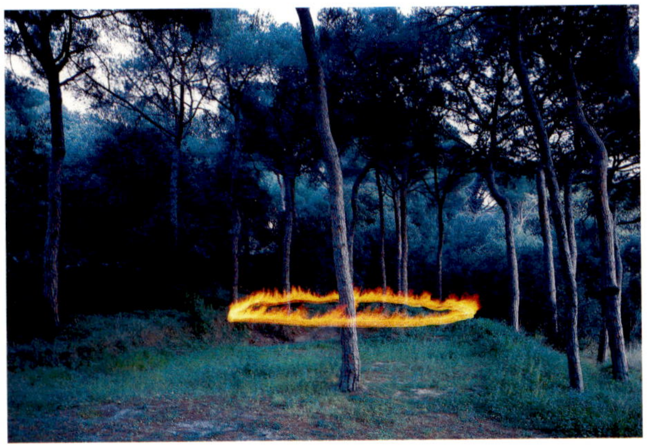

Fertility Ring. Allegory of Celtic Rites. "Fire" series
Tiana, Barcelona, 1986
Cibachrome 15,7 x 21,6 in

When his convalescence was over, he left the Casa Mila with not only a different attitude to his work but also a new passion for architectural photography, particularly for photographing fantastic architecture. Over the next fifteen years, his photographs would appear in several architectural publications such as, Casa *Vogue* and *db Architecture*, as well as in a publication on Sephardic sites in Spain, *Sepharad-Sepharad* and a book on the Casa Mila roof terrace *El Jardín de los Guerreros*, (The Garden of the Warriors).

But it was a new group of photographs, more vital in nature, which he had begun working on simultaneously that he found the most intriguing and fulfilling. The "Elements" series marks a defining moment for Armengol and from this point on his work references the most essential, fundamental components of our world. Rich in symbolism, it is an almanac of timeless images imbued with magic, mystery, art, nature and humanity.

Nothing is more fundamental to life than the elements and of these none has fascinated man more than fire. So, that is where he had started – turning to the literature on the subject he read the works of Gaston Bachelard and Sir James Frazer on the mythology & psychology of fire. Initially he made oil paintings of the powerful images these readings evoked because he was unable to find the images to photograph that would adequately express the complexity he wanted to portray. He realized that to truly accomplish that he would need to create the environments he envisioned and photograph those.

He began to seek out suitable sites, roaming the countryside and seashore until he came upon an appropriate location and once he found it creating the exact situation in which to ignite the flame. The ensuing body of work is a sequence of ethereal, primordial images of fire – bathed in an eerie blue light a roaring inferno fills the mouth of a cave, a quivering flame glimmers at the edge of an ocean and an undulating, blazing, suspended circle that dances about the trunk of a tree.

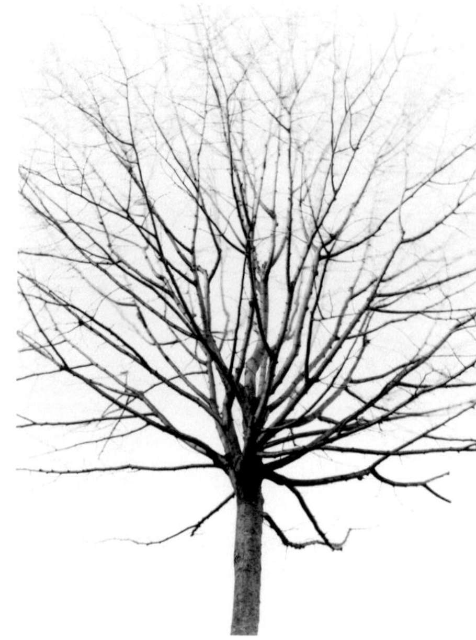

To continue with the "Elements" series, he moved on to wind (air) and water. As he experimented with different forms of expression he was also looking for new ways of making the photographs. He acquired an antique large format camera, decided to concentrate on the purity of the black and white image and to go back to printing his own photographs, which he had not done for years. He befriended the Japanese architect and poet Elia Taniguchi and together they produced two exquisite artist's books in 2001, *Voices of Water* and *Memories of Winds*, with Taniguchi's poems and Armengol's photographs.

Concurrently with "Elements" he began work on a piece that addresses an equally essential aspect of our existence. The alphabet is the basis of the word, of language, of what makes us human, it is symbol, sign and signifier. Armengol's "Alphabet Signs" arose from his conviction that in our globalized world the current social, ethical system is breaking down and must give way to a new set of values and human interaction. This breakdown, he believes, indicates that humanity's survival is in jeopardy unless new parameters for a peaceful, harmonious coexistence are devised.

The Linden Tree. "Elements" series
Tiana, Barcelona, 1995
Gelatin silver print, 23,6 x 19,6 in

The Linden Leaf. "Elements" series
Tiana, Barcelona, 2001
Gelatin silver print, 23,6 x 19,6 in

To achieve this, he is convinced we have to go back to zero and invent a different language of communication based

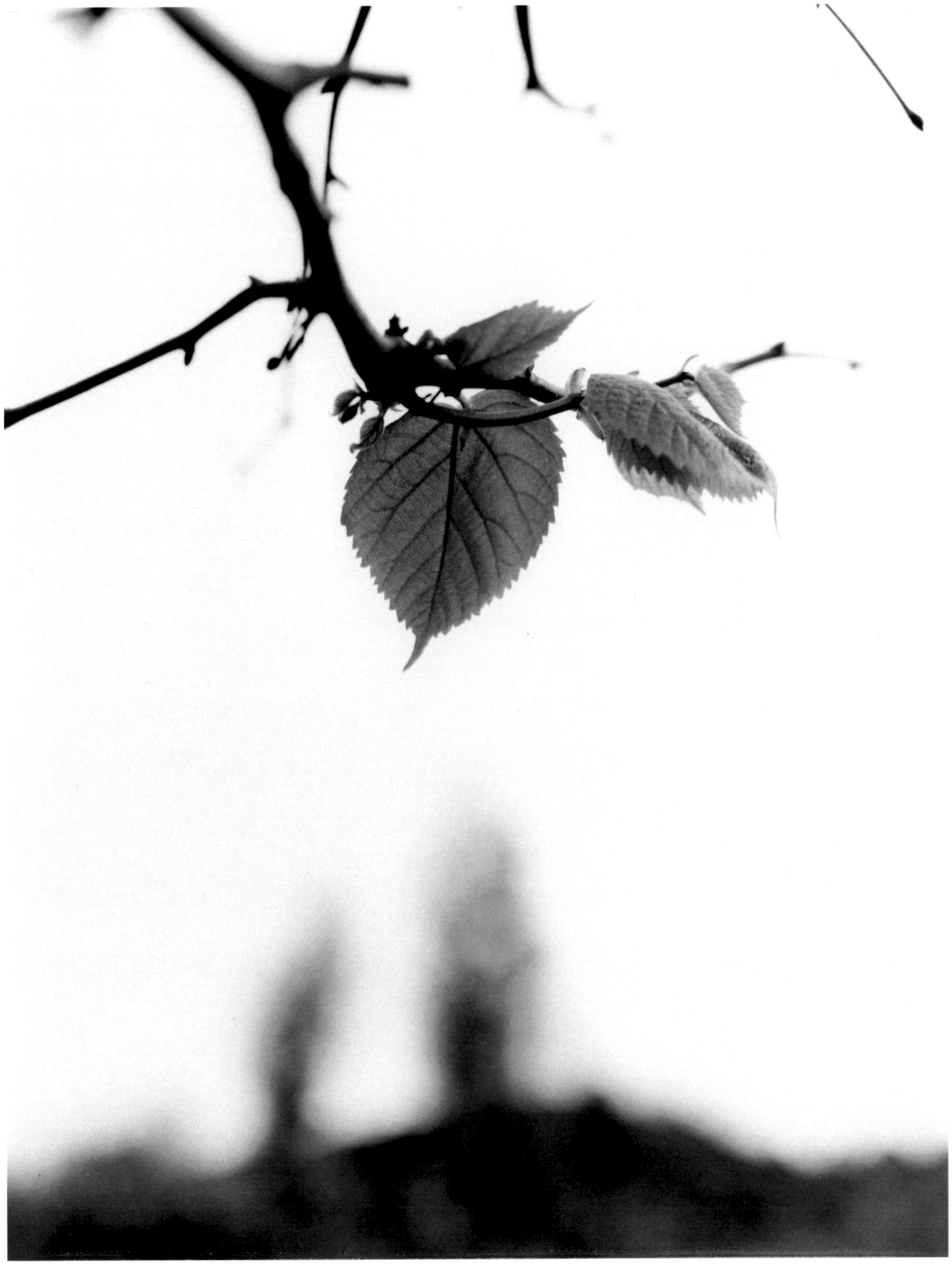

on a new "alphabet" – a system of signs that could express more fully, more philosophically what all mankind has in common.

"This piece brings together several of my interests, sculpture, photography and symbolism. It consists of 22 photographs, the number of cards in the Tarot, which is inspired by the Hebrew alphabet, and uses images, numbers and symbols to indicate values, attitudes and situations of the individual in the world."[1]

He believes this reinvented "alphabet" should emerge from the world of nature, the one force that unites us all, and to create "Alphabet Signs" he turned to the fourteen olive trees on the small piece of land he inherited from his father. Using the branches of the olive trees, a symbolic gesture in itself, he selected those whose lines, or movement reflected what he wanted to convey and photographed them in high contrast so they appear bone white against their black background.

According to Armengol the black background he utilizes in all of the alphabet images refers to the genesis of time, to the darkness before light. They speak to us of the allegory of space and time, about the mystery of life and our underlying quest for its essential values. Their lines, enigmatic strokes from some arcane pen, black empty spaces that trap a tiny gesture, hint at directions that could be taken. The work can also be viewed as a kind of game – one's own values can be linked to the representations, in the form of a symbolic language, as though they were fundamental in establishing the graphic expression of what would then become an "alphabet".

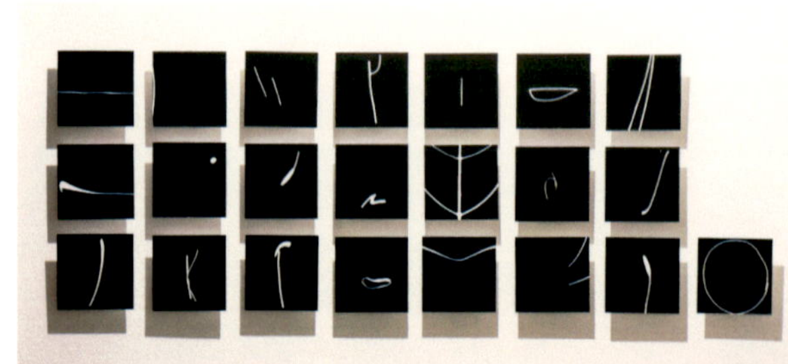

"Alphabet Signs", polyptych
Barcelona, 2000

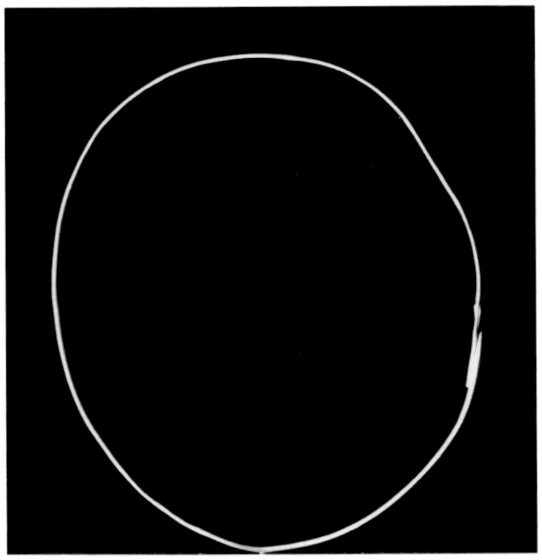

Zero. From the polyptych "Alphabet Signs"
Barcelona, 2000
Gelatin silver print, 8 x 8 in

"Alphabet Signs" may be informed by the Tarot, but it is overwhelmingly an alphabet for our time. Despite its esoteric symbolism, there is much of the human condition in it. In addition to being a reflection on our existence in these unsettling times, it contains fragility, a hint of pathos and a certain fallibility that is all too human.

The piece is actually comprised of 21+1 photos-signs; the extra one being the *Zero* which is usually exhibited more prominently than the others, as it should be, for the zero has powerful connotations and here against a void-black backdrop it speaks volumes. However, Armengol's pared down *Zero* is slightly wobbly, somewhat frail and frayed at the edges. In fact, it looks like it might well have stood up from a page in one of Samuel Beckett's plays and would not look altogether out of place rolling along that country road in *Waiting for Godot.*

The reference to Beckett is relevant for there are elements in much of Armengol's work that echo Beckett in his paring away of the superfluous from his lexicon of images to expose only the essential. His work is deceptively simple, devoid of artifice and the materials he utilizes extremely sparse.

Since the material used in the photographs that make up "Alphabet Signs" is so sculptural it is not surprising that working on this piece led him to an ongoing project comprised of a group of sculptures made from branches of the same olive trees. In making the sculptures, Armengol says he searches for certain "expressions", lines or gestures in a particular branch that might appeal to him, such as a figure with hands aloft beseeching help from above, then he fashions a base from a small piece of slate left over from the roof of the simple wooden house he has built on the terrain. Some of these sculptures were later cast in bronze and included in an exhibit together with "Alphabet Signs".

Armengol strives to work in harmony with what could be referred to paradoxically as "slow" time, or as he calls it "long" time – working in conjunction with time, not in opposition to it as most of us attempt to. In the making of his photographs he works with time measured in natural light under whose dictates he waits for the moment, the right "time/light" in which to pull the shutter. Light is a fundamental component in his work, and of enormous importance, as is reflected in a sequence of photographs he made on a trip to Iceland in 2003. The extraordinary light in these images is not the light of day; the sensual light born of the sun, of warmth, rather it is a hard, cold virginal light that penetrates night.

"Chronicle of Iceland" points not only to a mythological past but perhaps also refers to an all too real future, one predicted by global warming. In these dramatic, stark images we catch a glimpse of our planet as it was in prehistory, locked in the grip of the elements, clenched by ice. Great rivers of white water cut through coal black earth and sheets of luminous ice glow against saturated slush-grey clouds. Decapitated volcanoes loom on bleak horizons in a world bereft of the trappings of our era and devoid of any plant life.

It is to precisely this absence that Armengol turns in his next body of work, "Herbarium". In the 118 photographs that comprise "Herbarium", 88 of which are included in this book, he uses the camera to celebrate nature in one of its most generous, simple and elegant forms.

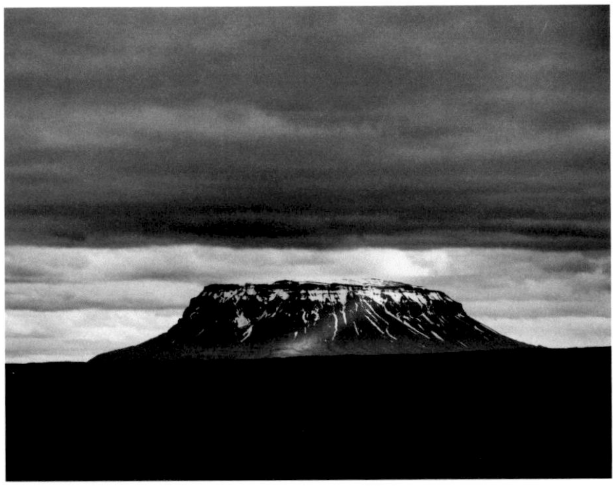

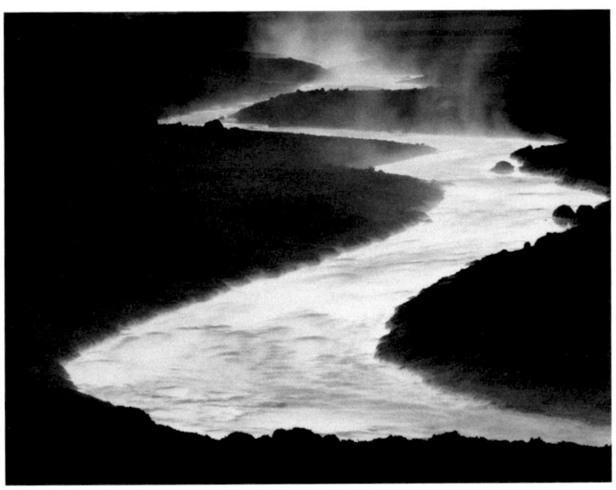

Blafjall
Iceland, 2003
Gelatin silver print, 15.7 x 19.6 in

Krafla
Iceland, 2003
Gelatin silver print, 15.7 x 19.6 in

The traditional Herbarium with its magical and curative associations is deeply rooted in our collective memory and is as ancient as our desire to name and give order to the natural world we inhabit. Many of the plants depicted here are among the oldest on the planet and have had their present form since time immemorial. Usually only their size has fluctuated over the centuries as a result of changes in Earth's atmosphere and climate.

"I've been in touch with nature since I was a child, Armengol explains, at times more frequently and more closely than others. However, at the beginning of the '90s I felt the urge to surround myself with it, to contemplate it, and acknowledge it through photography. It came upon me almost stealthily, without my really knowing where and how to begin."

In creating the "Herbarium" series he traveled some seventy miles from Barcelona to a meadow on a friend's property in the Alto Ampurdan region during a four season cycle that lasted from spring 2004 until winter 2005. On these trips, he chose a section of meadow where he concentrated on a group of plants of similar genus growing in what he refers to as a "little berth of nature".

Selecting about seven of the plants at a time, he photographed them with a 35mm camera before taking them out of the ground. He then wrapped the roots and stems in a damp cloth and rushed back to his studio to photograph them with his large format camera working in sessions that could last for several days depending on the state of health of the plant. To him plants are living entities with myriad characteristics that exist unperceived beneath our jaded gaze, silently succumbing to the machines of our civilization. His objective with this body of work was to make these simple, overlooked herbs the focus of our attention.

According to art critic Donald Kuspit it is an aesthetic feat to see what is in front of our eyes. "It seems beside the point of our everyday relationship with things to see them aesthetically, never mind ecstatically – it is easier and more practical to deal with them as matter of fact signs of themselves... to deal with them otherwise can be emotionally dangerous", he writes in discussing the work of Albert Renger Patszch, whose photographs challenged widely held perceptions of beauty, which for the great German photographer was resplendent in "everyday" things.[2]

Armengol's intent in creating "Herbarium" was to use the camera to rescue these seemingly simple plants from anonymity and show their serene beauty. His *Herbarium* does much more than that. In these photographs a metamorphoses takes place whereby the image of an unpretentious herb is transmuted. The commonplace becomes profound and mysterious, transformed into a banquet for the eyes. Vibrating with energy, leaves and tendrils swirl and curl in a gothic flamboyance, plump pistils perforate the air with sensuality; flower heads hang delicately on radiant stalks as the artist's vision enables us to see them in their sublime splendour.

As such, the plant form becomes also an art form; an intricate, infinite variation of lines, planes and volume, in a way of seeing facilitated not only by the intensifying experience of the camera but also by Armengol's obvious perceptual affinity with the plants he is photographing. Looking at his "Herbarium" images we marvel that we had no inkling of this magnificence as we casually tossed herbs into a soup, threw them aside while weeding or nonchalantly trampled them underfoot.

"It is the result of the creative act that distinguishes works of Art from those of Nature, namely, the modeling of an original form, newly produced, not the imitated, or repeated form", Karl Nierendorf wrote in his introduction to Karl Blossfeldt's renowned book of photographs of plant forms nearly a century ago.[3]

This maxim is evident, not only in Armengol's "Herbarium", but permeates his entire opus. His work is propelled by a force that prohibits a mimetic art, a force that is in constant pursuit of innovative forms that in a seamless melding of eye and camera brings to light the invisible, revealing the exquisite in the austere, the sublime in the simple and the magical in the mundane.

Margaret Hooks, October, 2006

[1] Quotations from Manel Armengol are from the author's conversations with him, Aug-Sep 2006.
[2] Albert Renger Patszch: "Joy before the object", Donald Kuspit, *Aperture*, New York, 1993.
[3] Karl Nierendorf, Intro., *Art Forms in Nature, Examples of the plant world photographed directly from nature by Karl Blossfeldt*, Berlin, 1928.

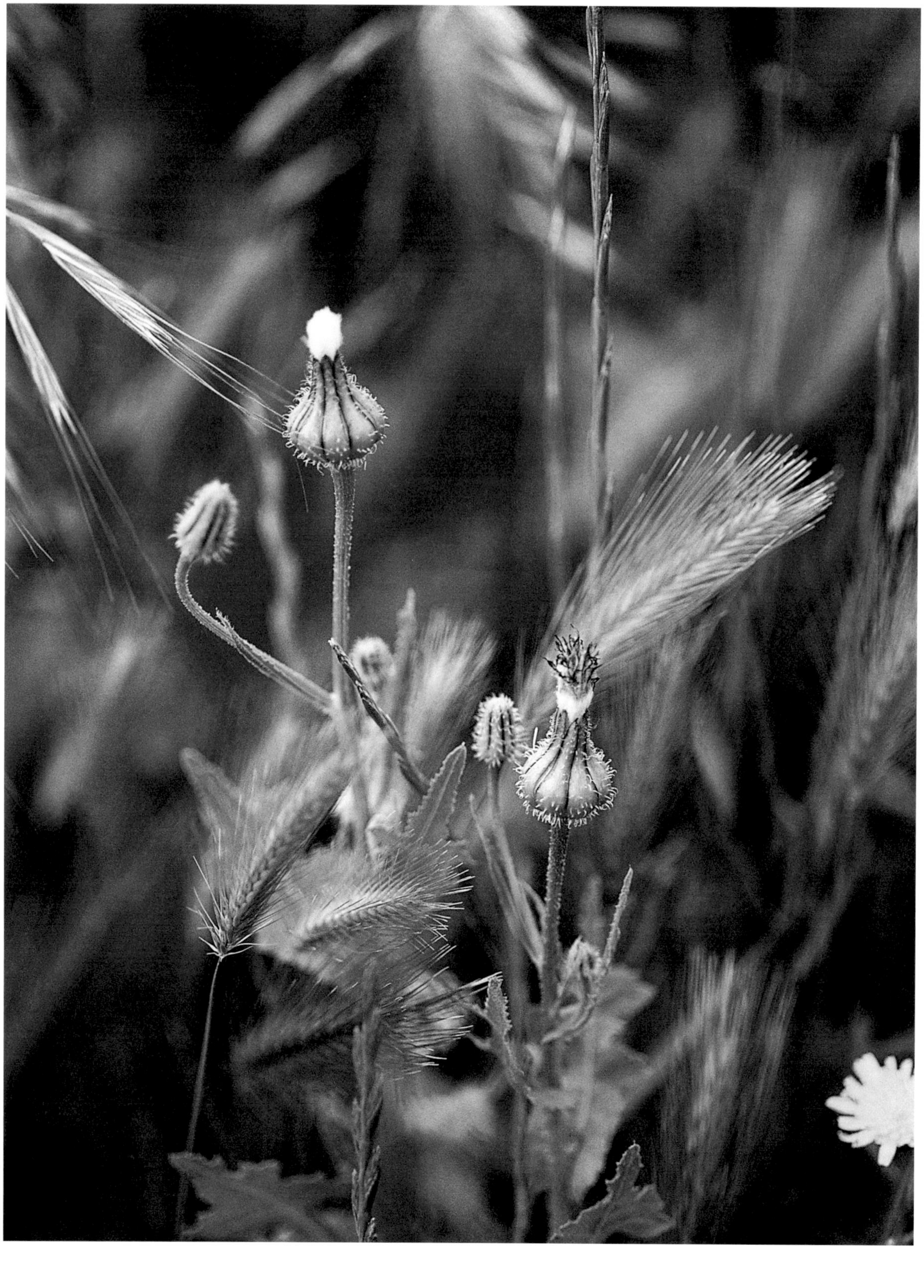

Bewildered
Amongst petals
Wings of birds

Masaoka Shiki

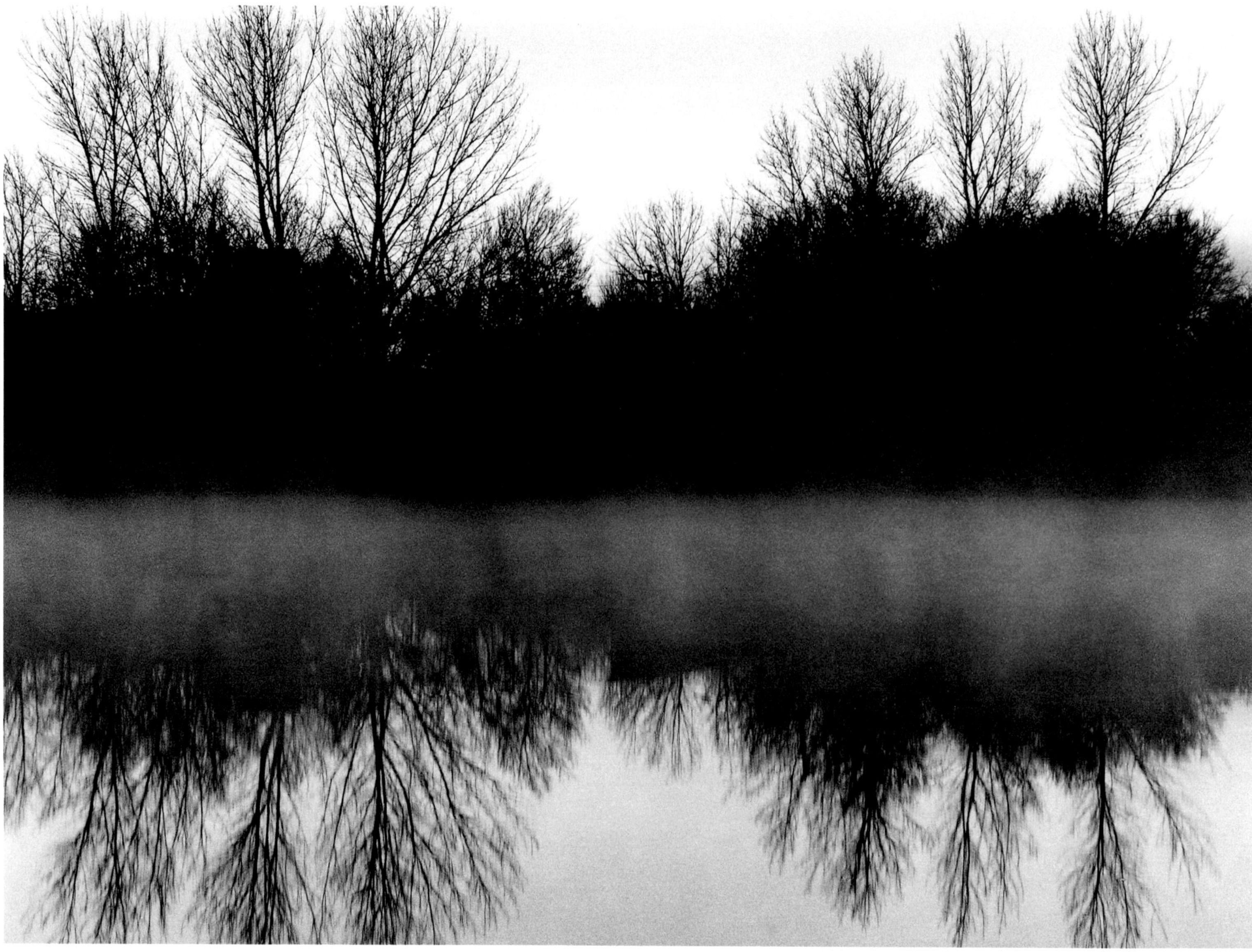

HERBARIUM

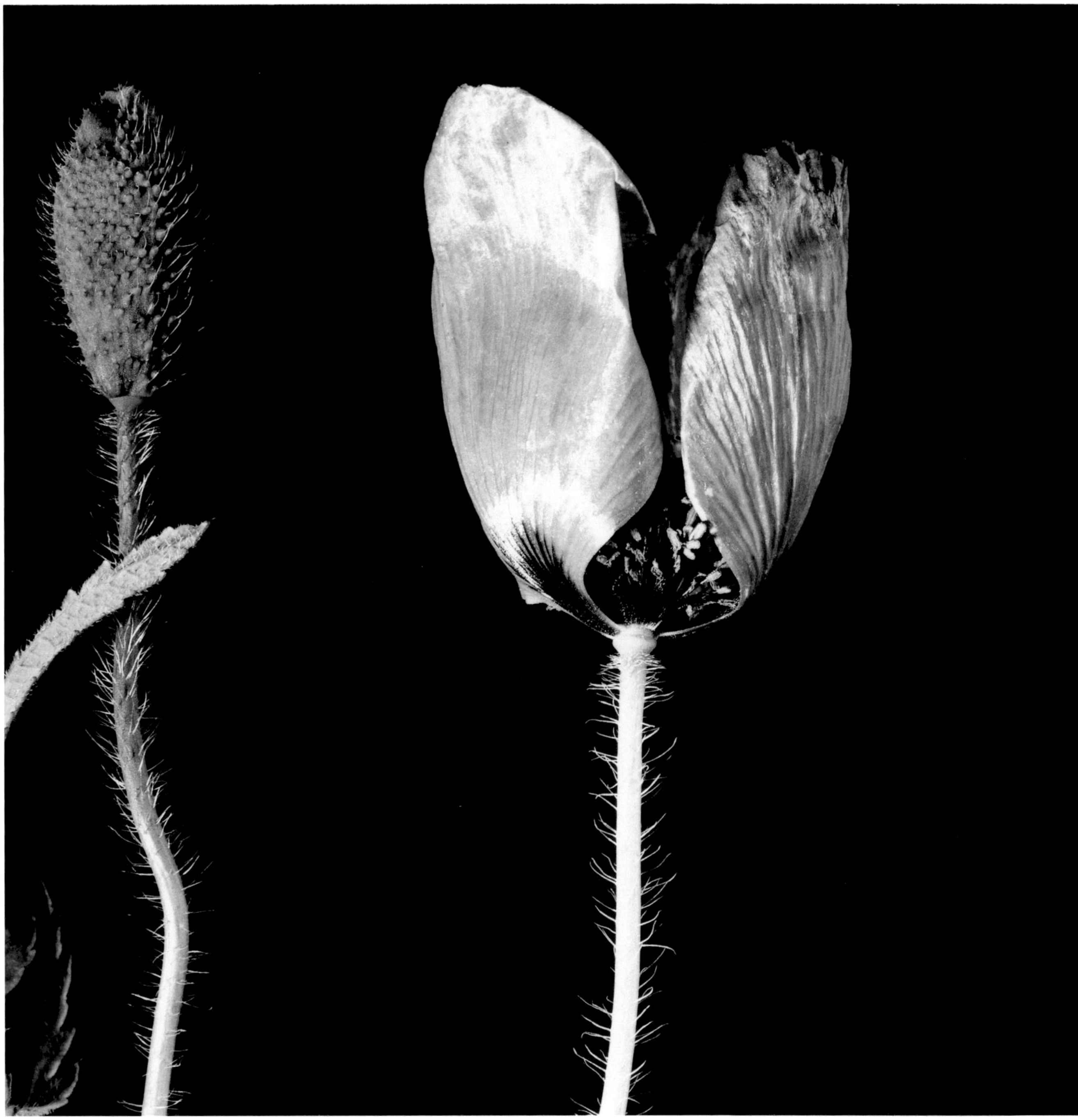

Papaver rhoeas

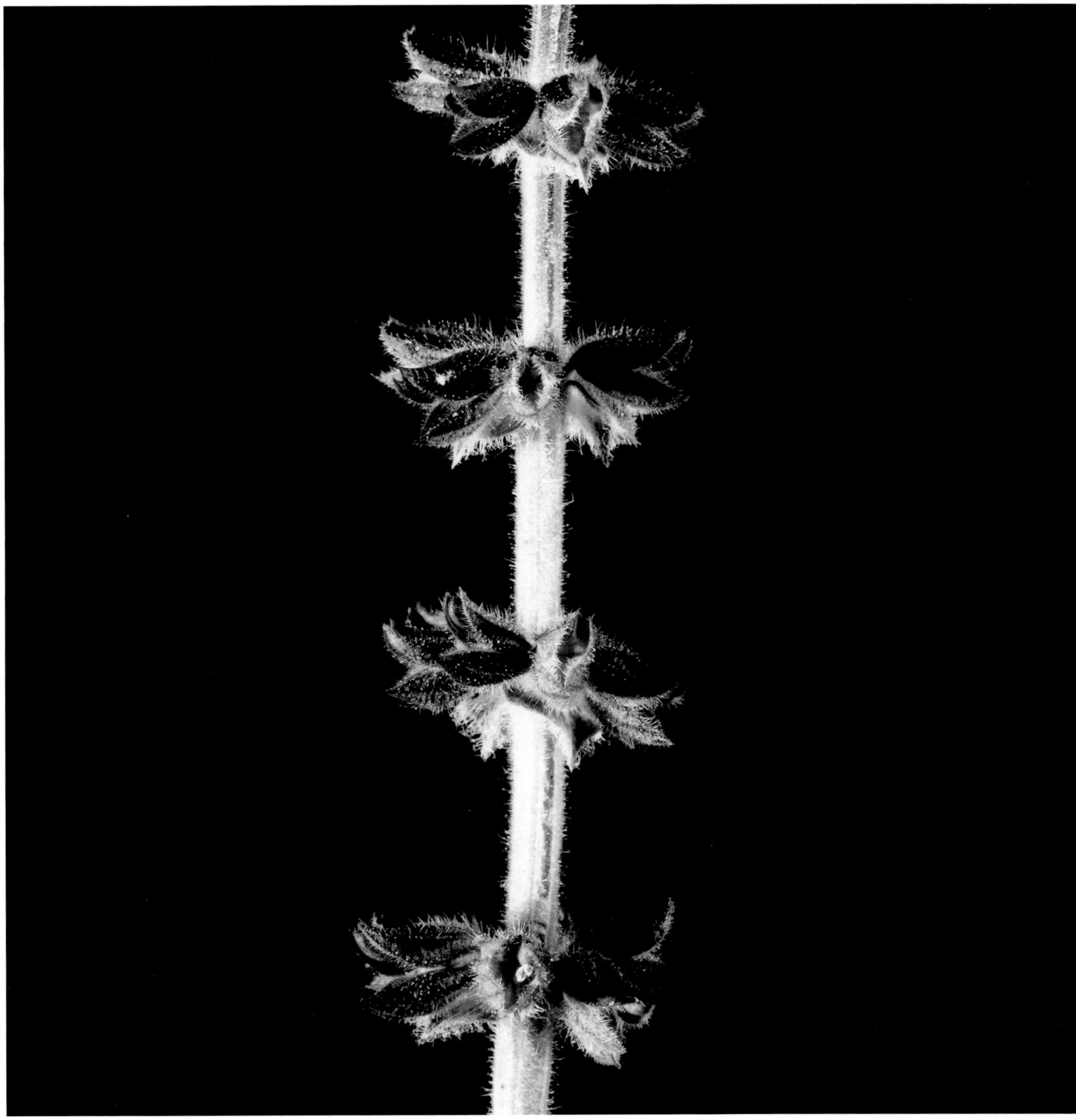

Salvia verbenaca

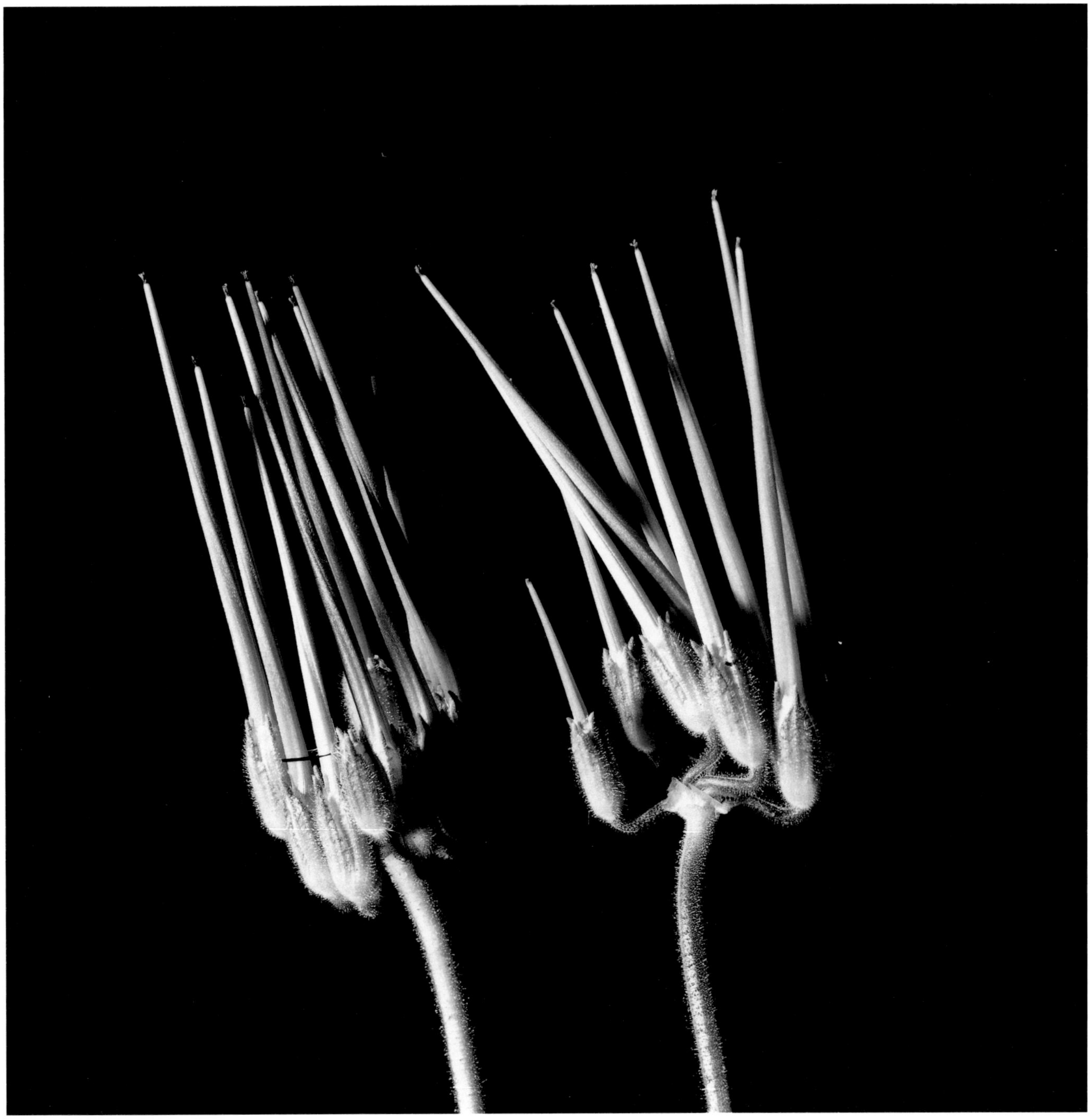

Erodium sp.

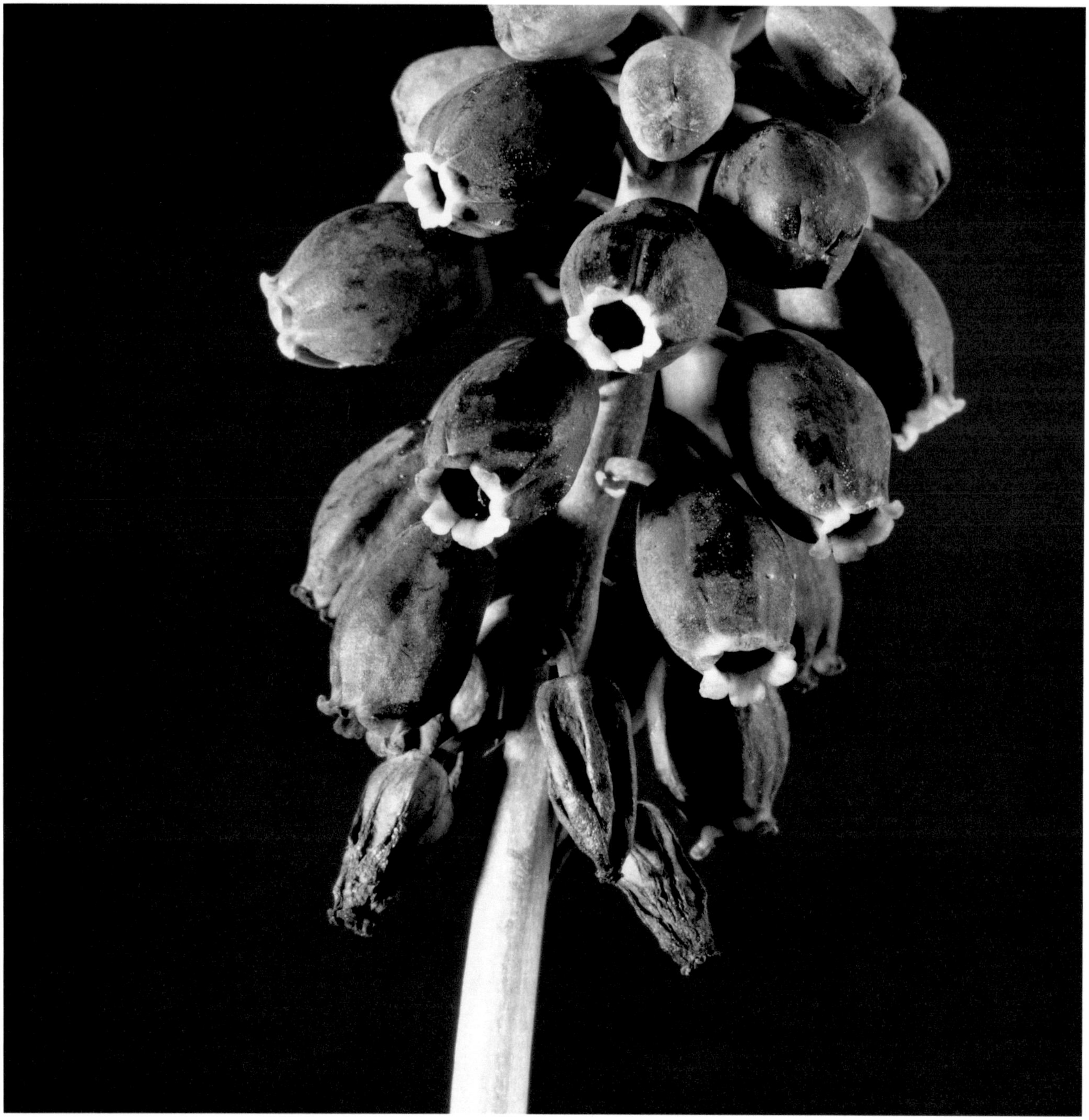

Muscari racemosum

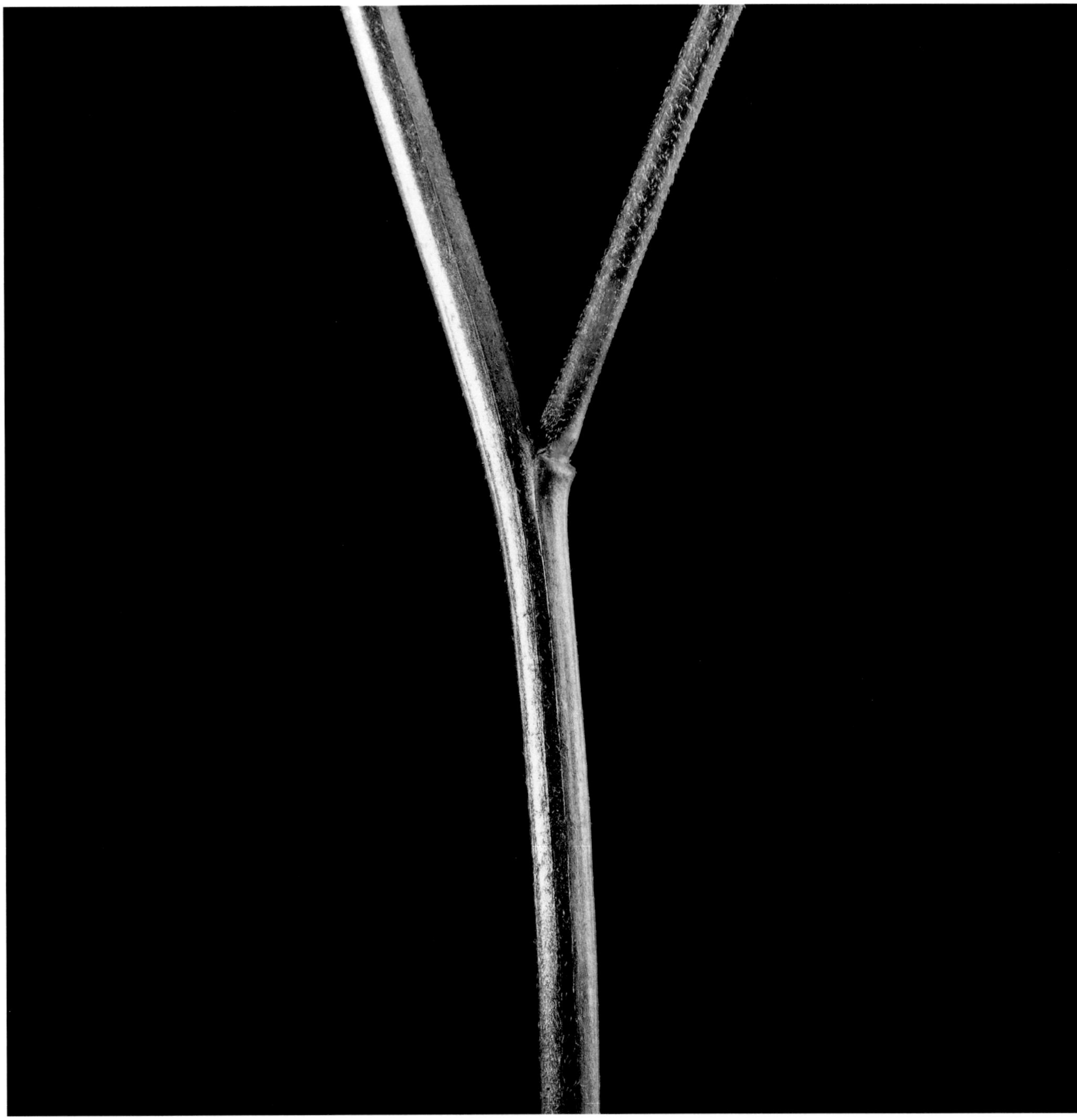

Crepis vesicaria subsp. *taraxacifolia*

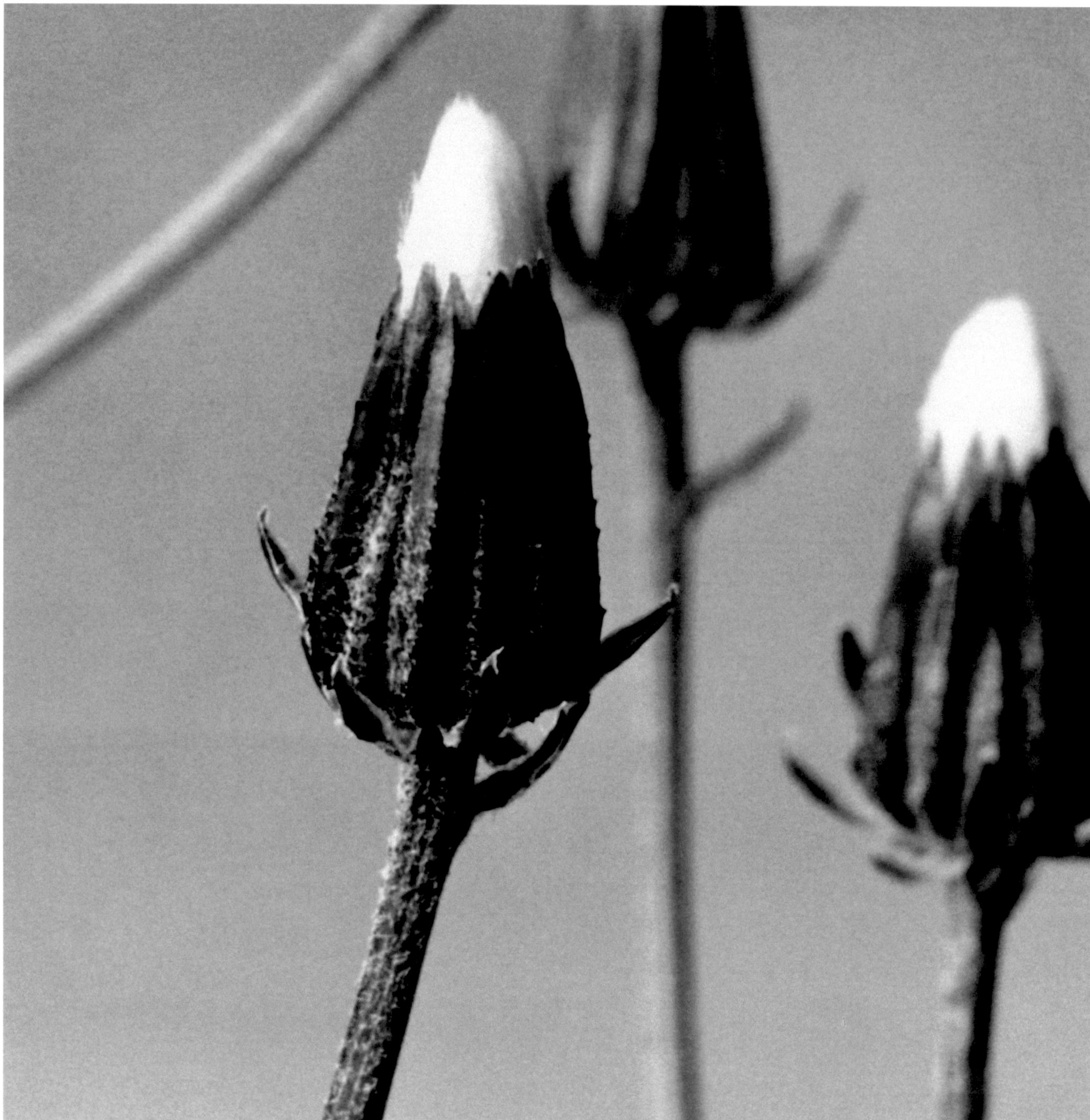

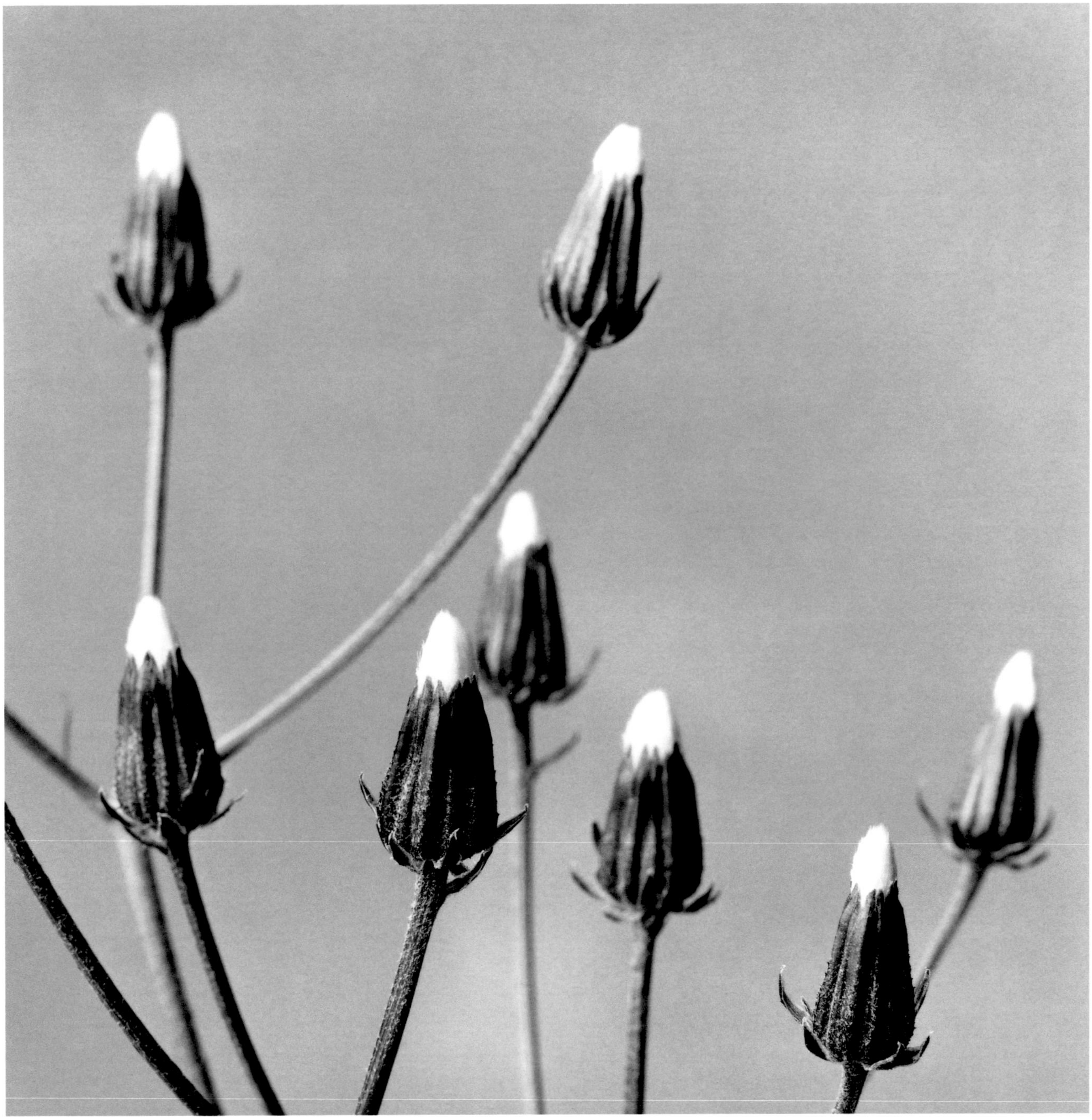

Crepis vesicaria subsp. *taraxacifolia*

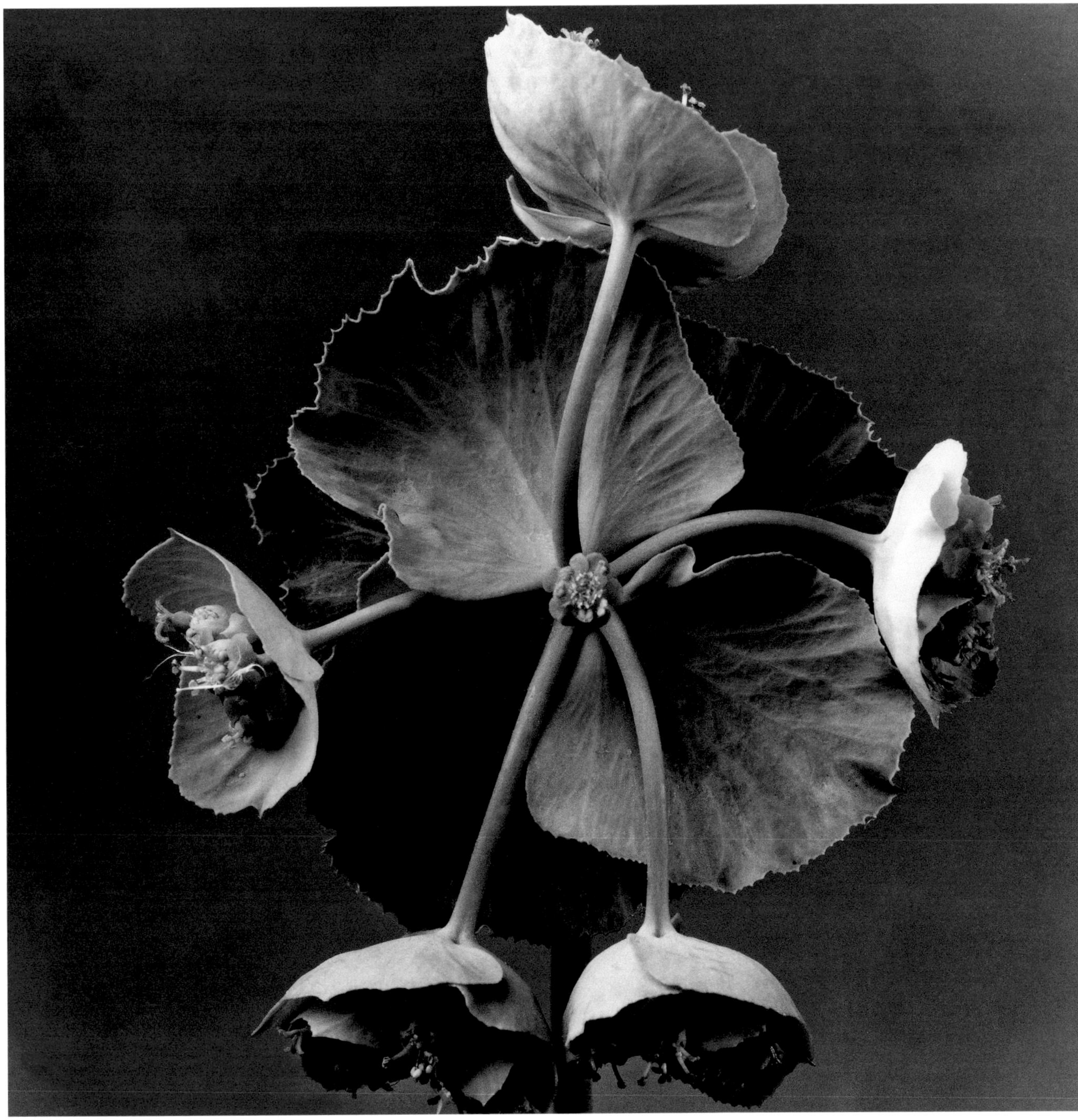

Euphorbia serrata

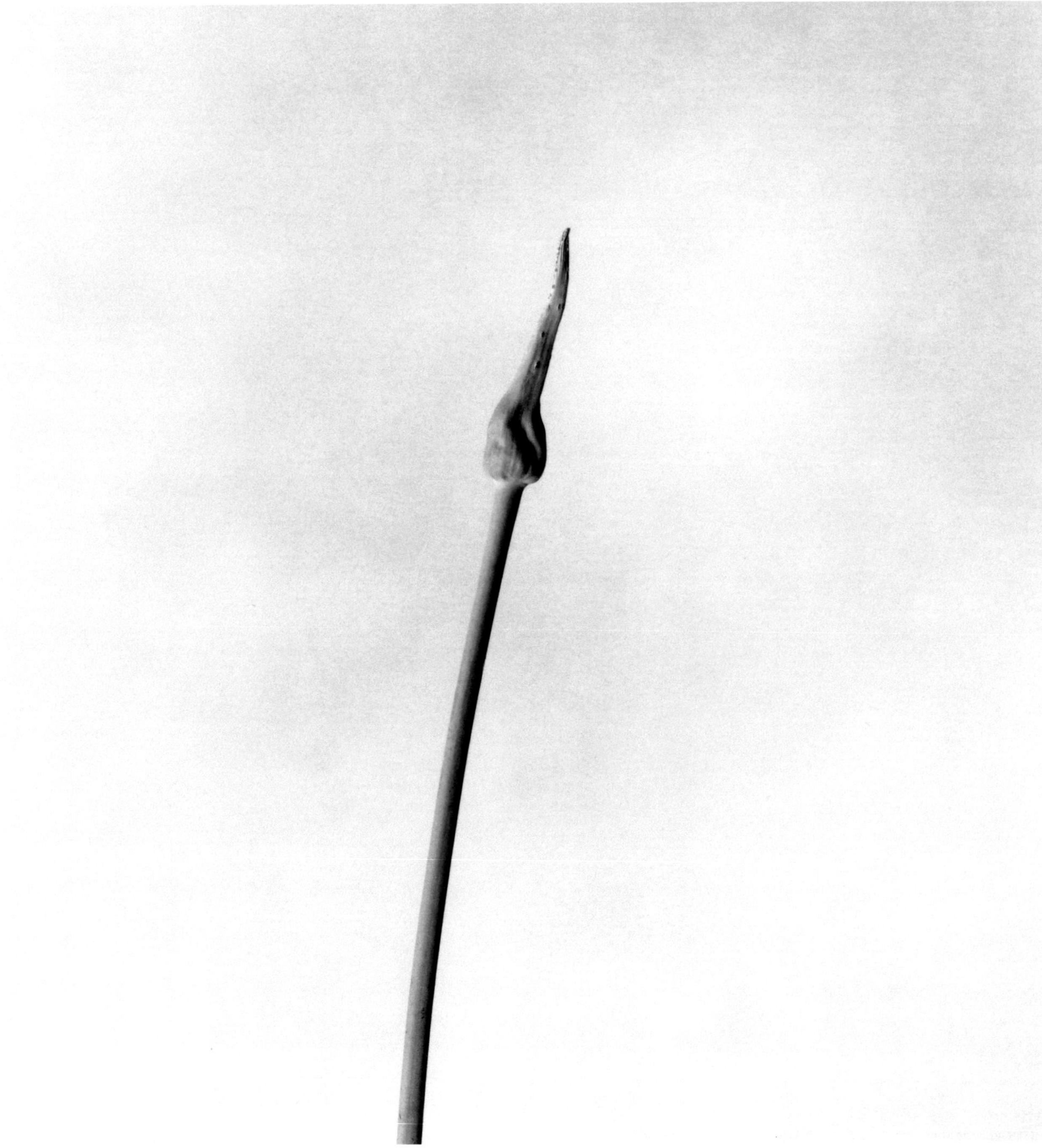

Allium ampeloprasum

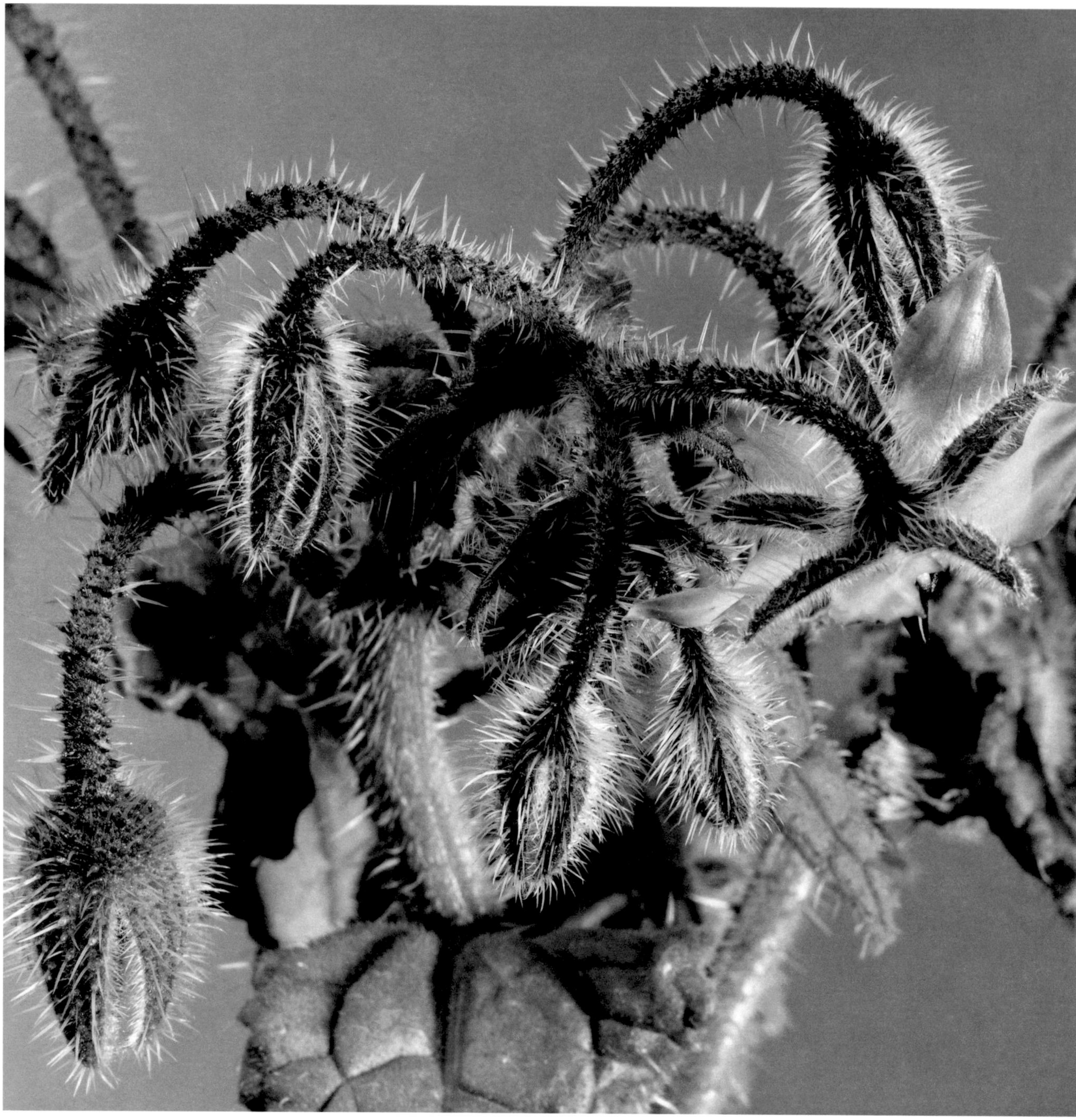

Borago officinalis

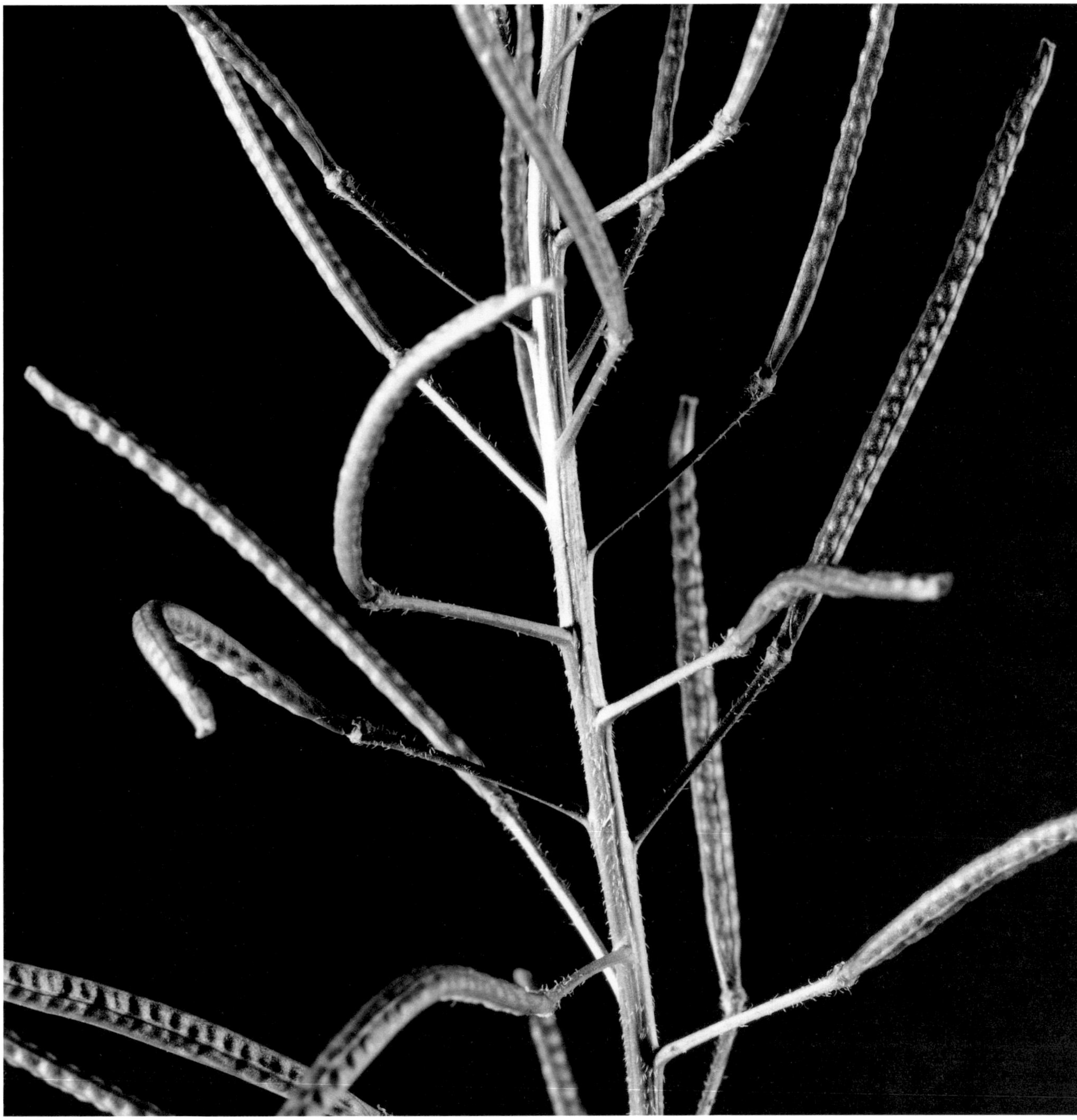

Diplotaxis erucoides

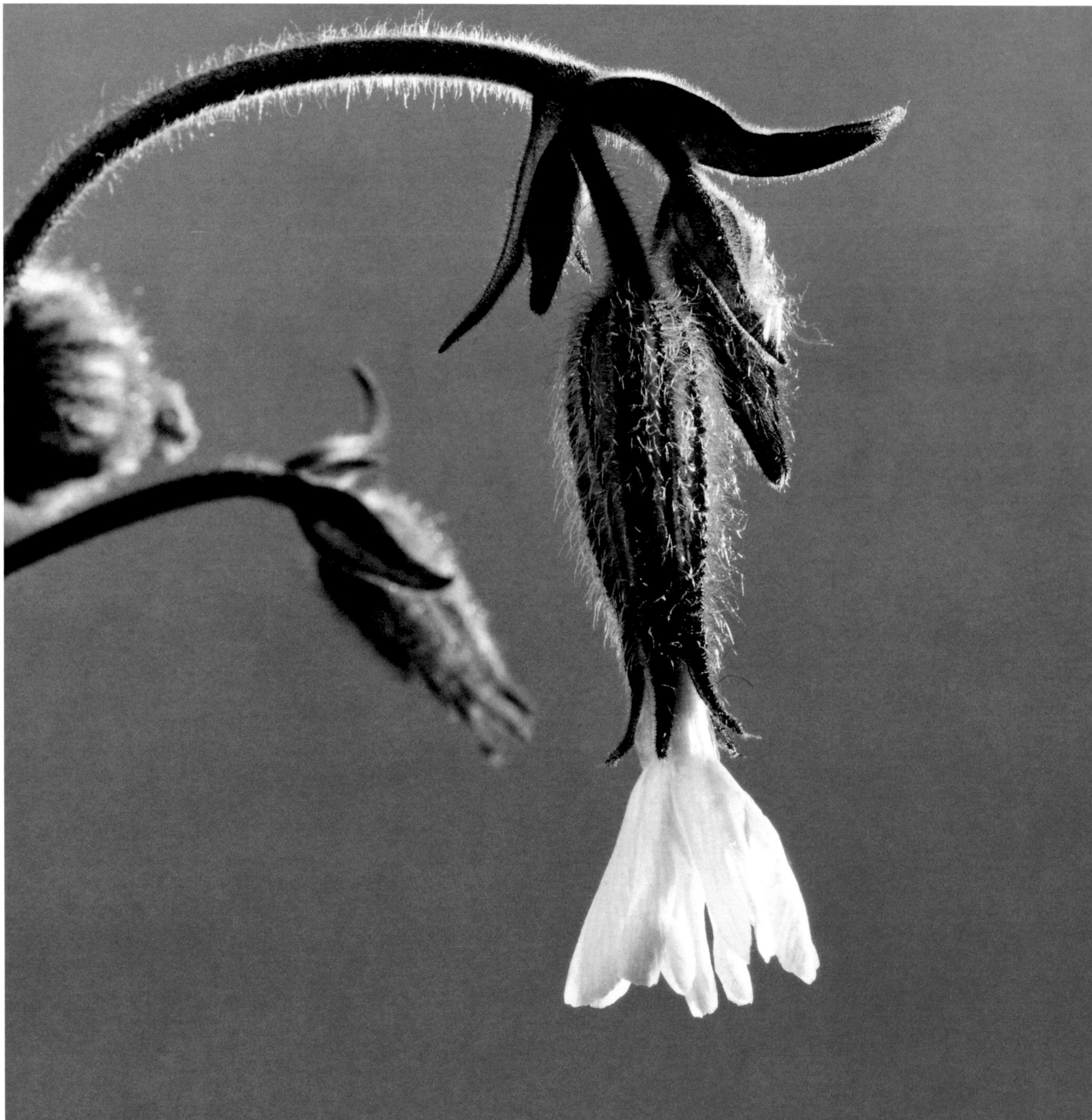

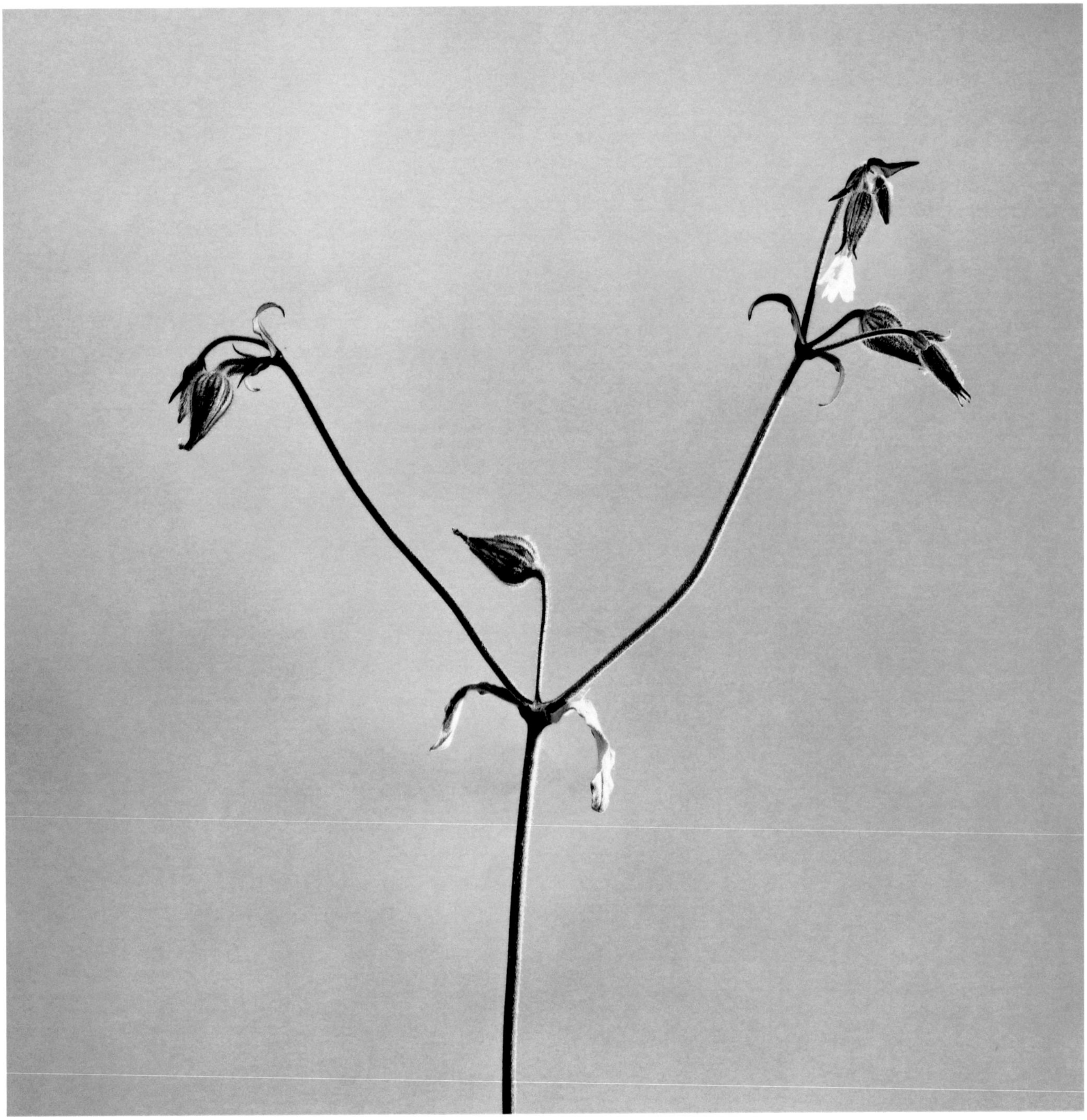

Silene latifolia

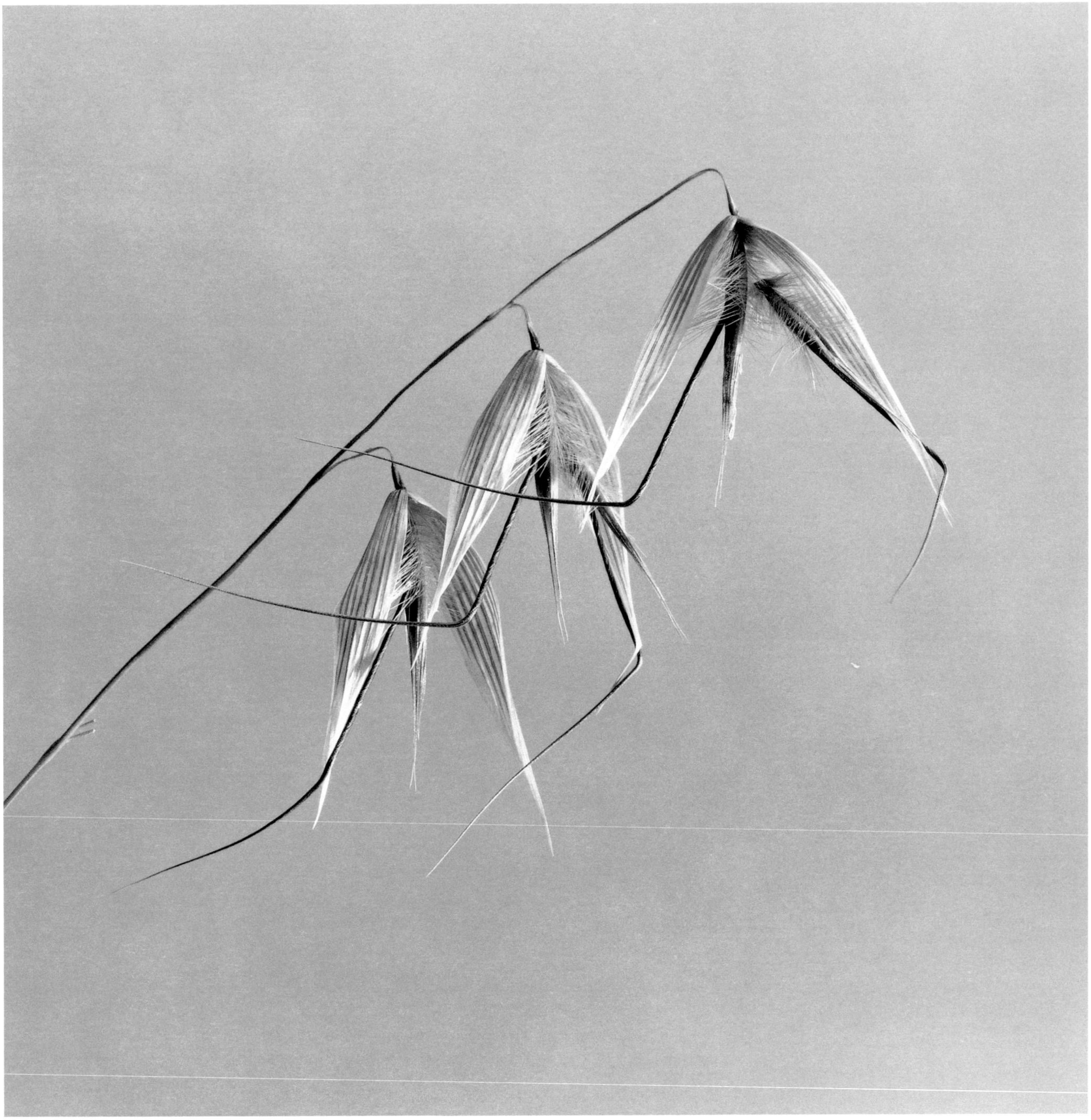

Avena barbata

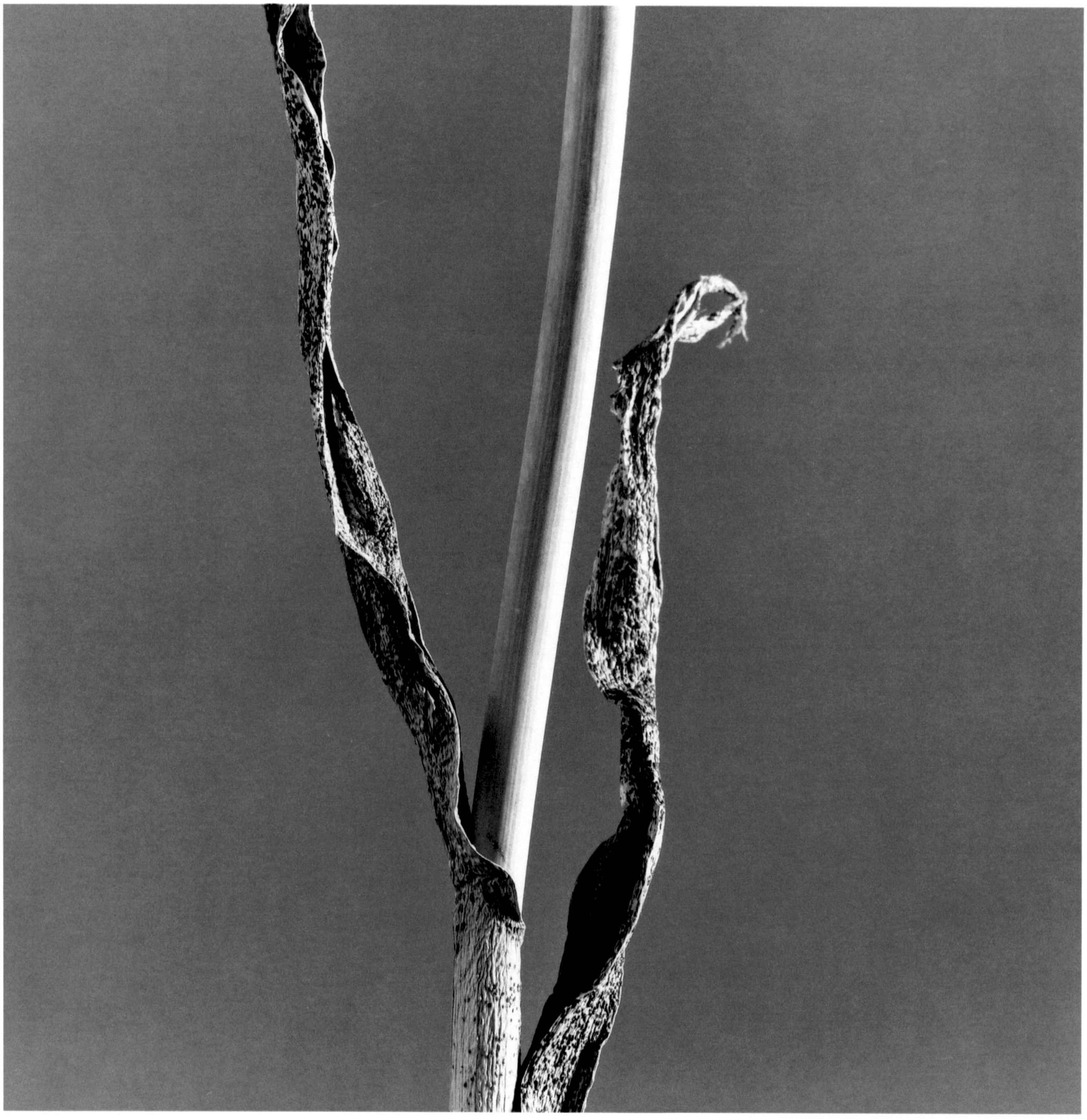

Allium ampeloprasum

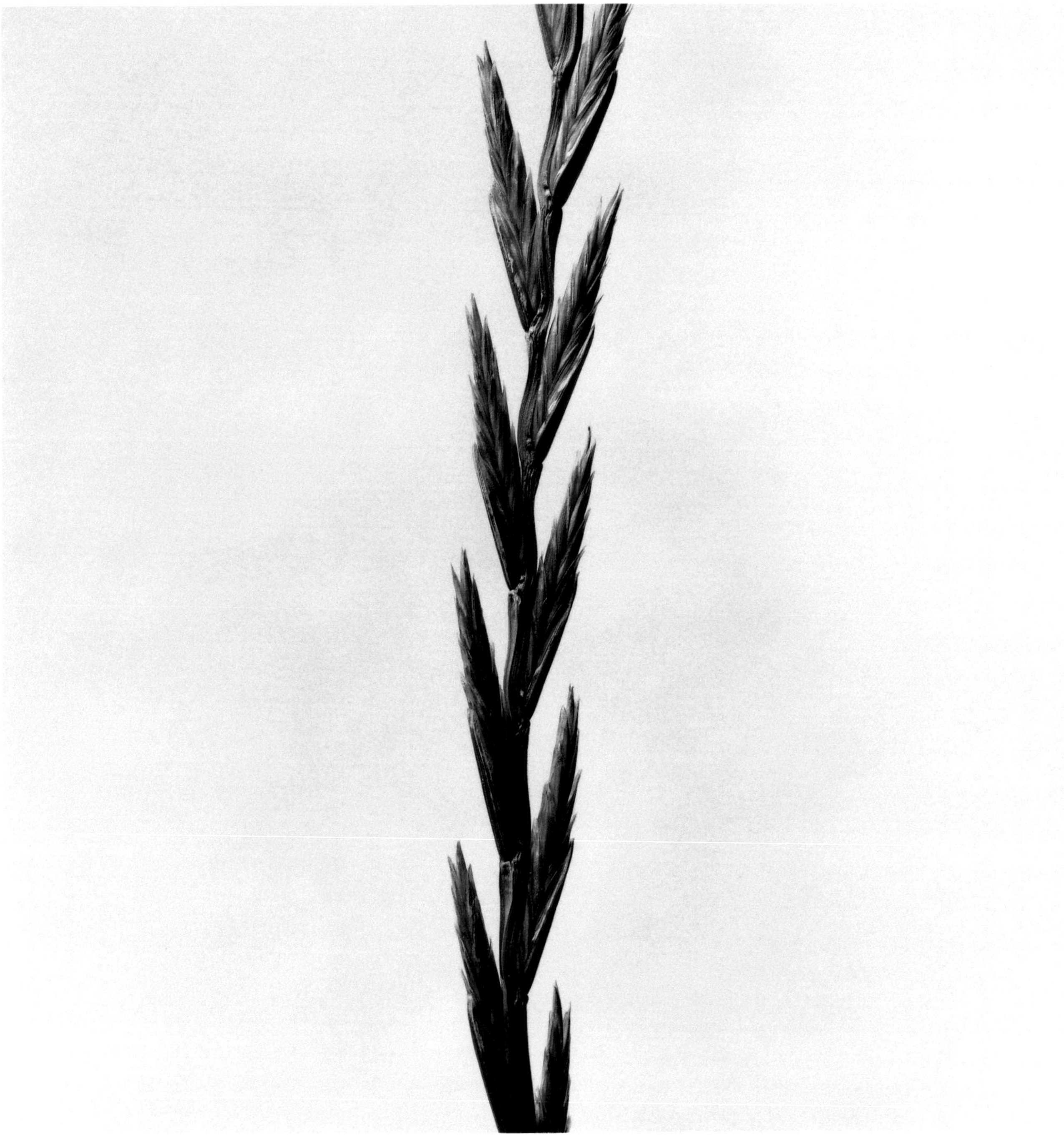

Lolium perenne

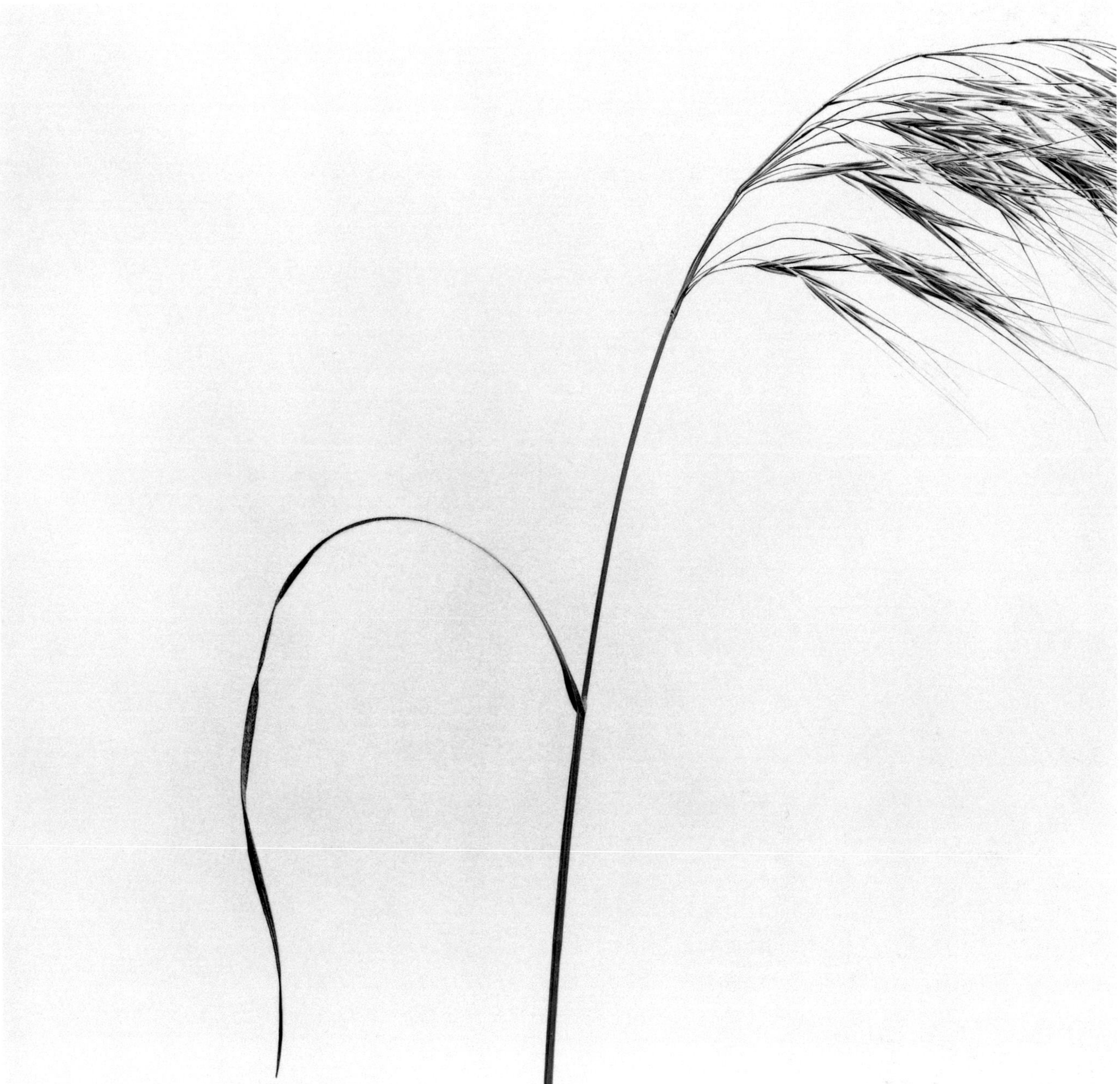

Bromus sterilis

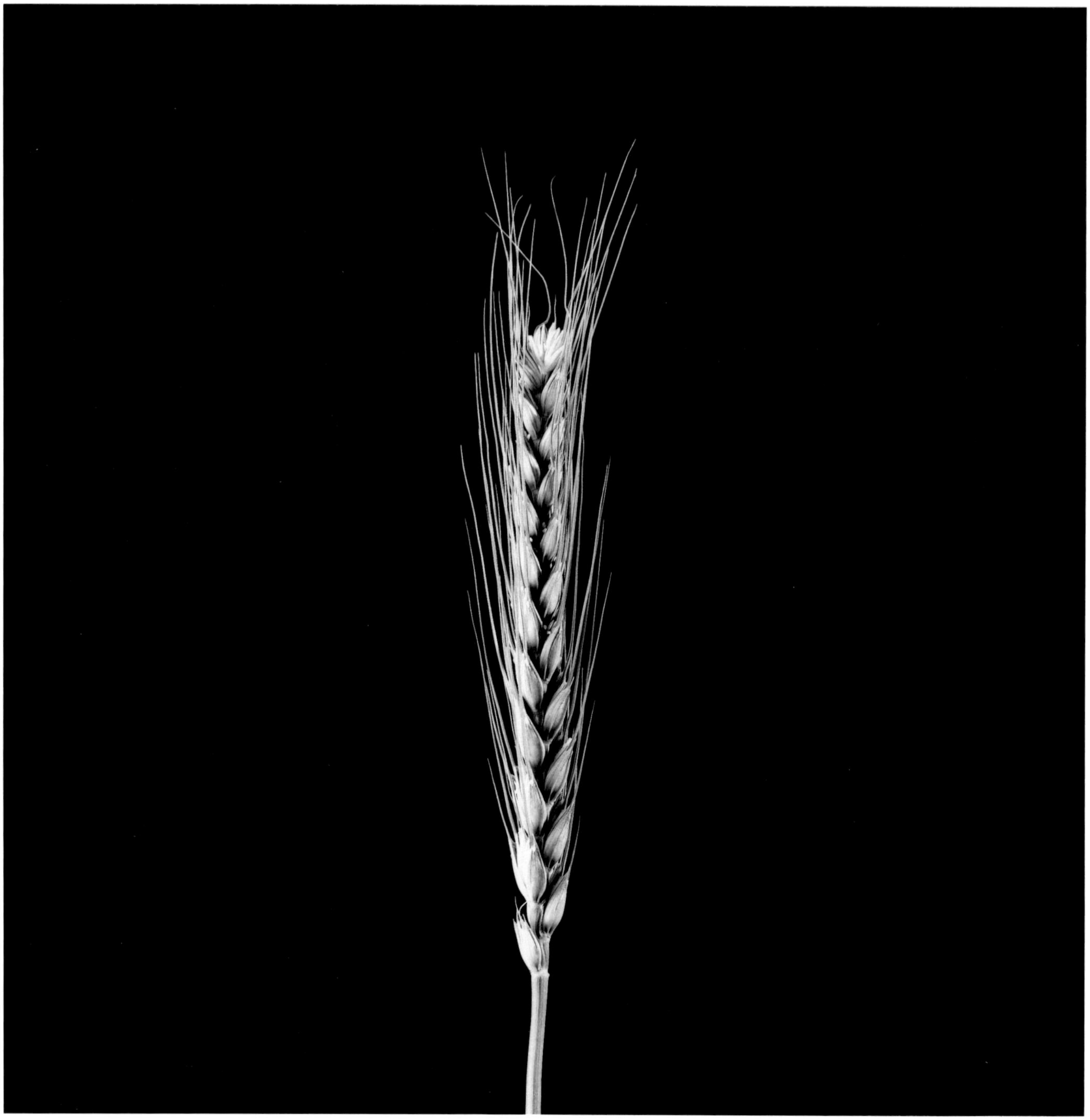

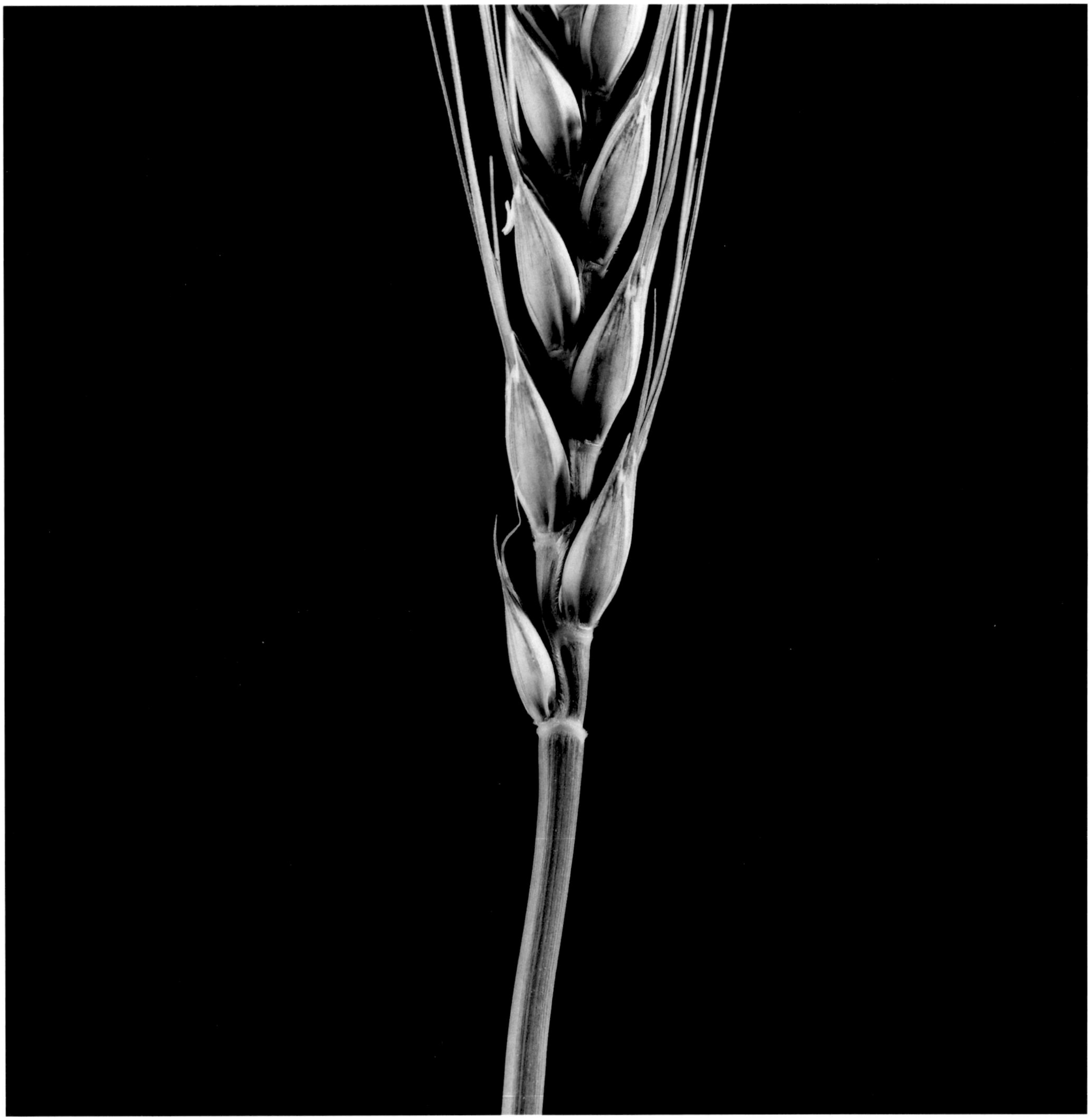

Triticum aestivum

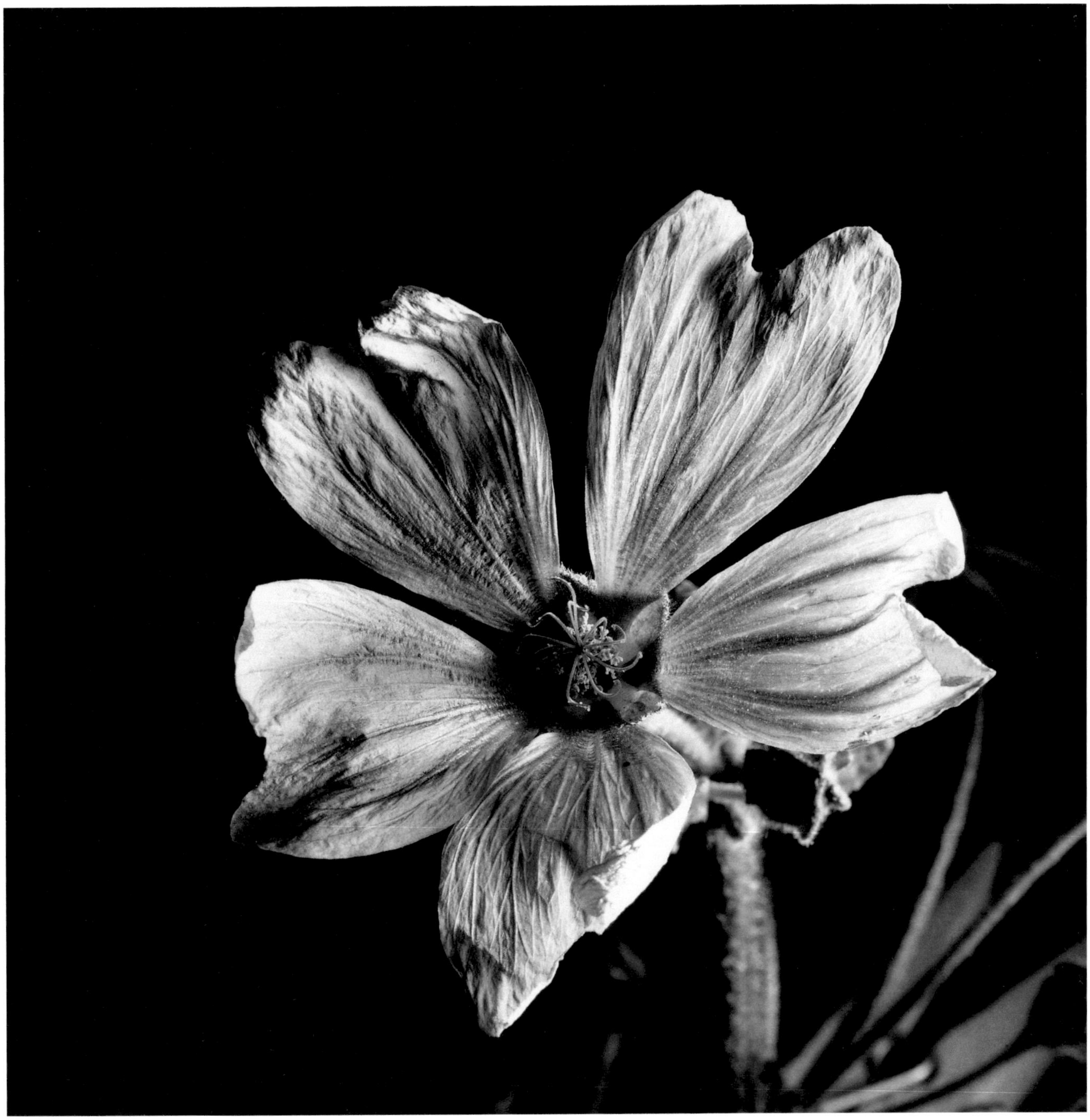

Malva sylvestris

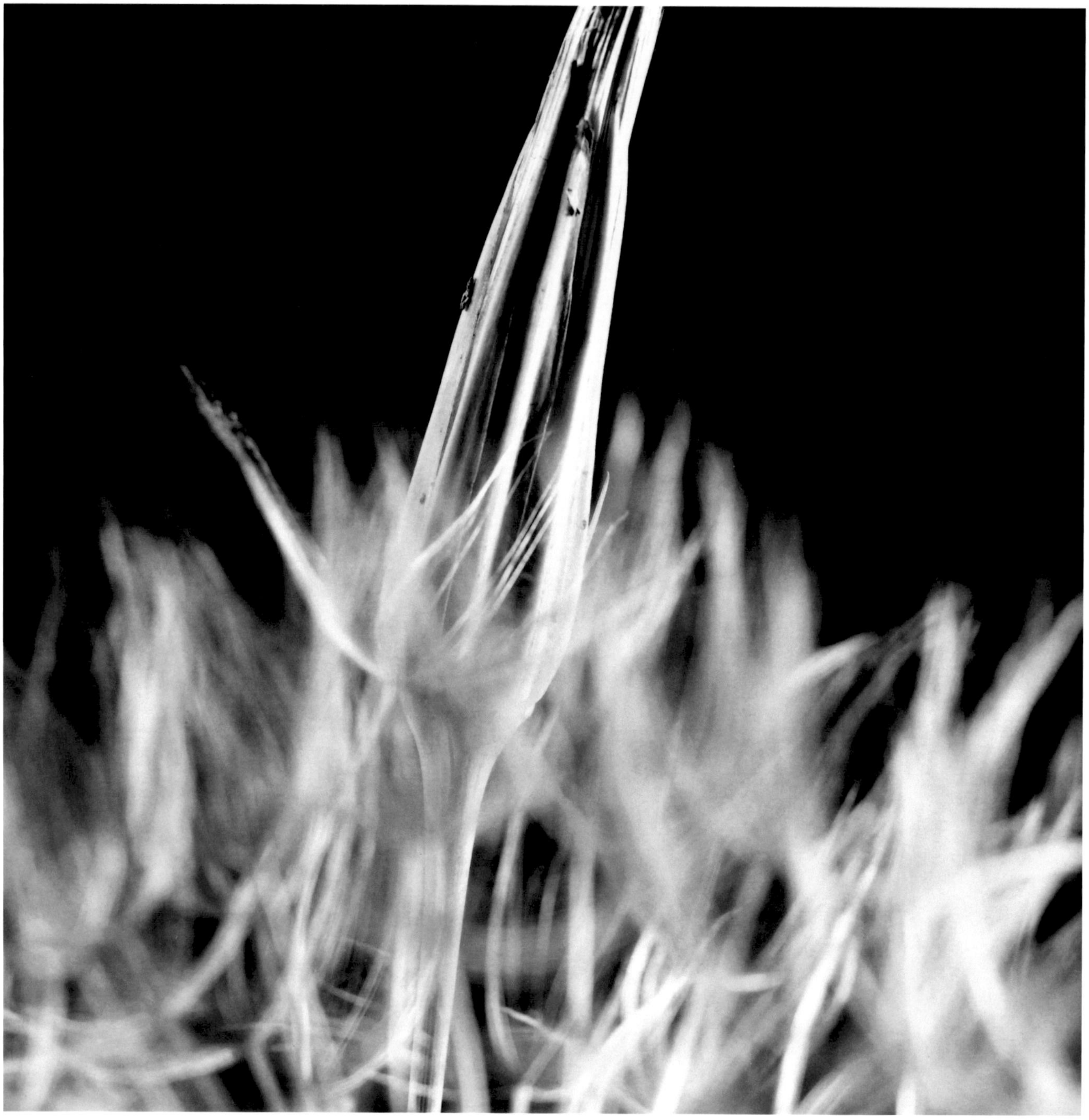

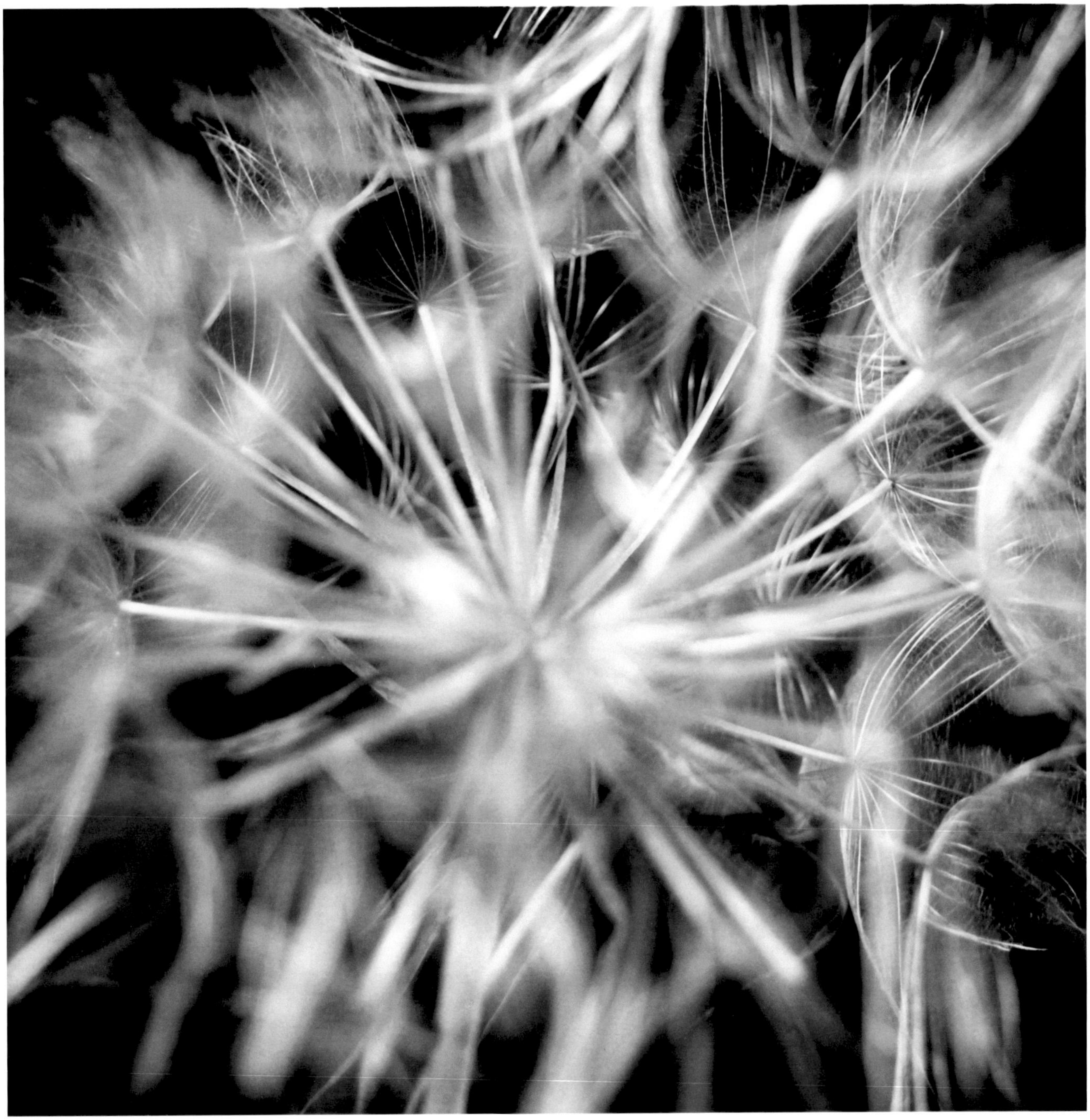

Scorzonera laciniata

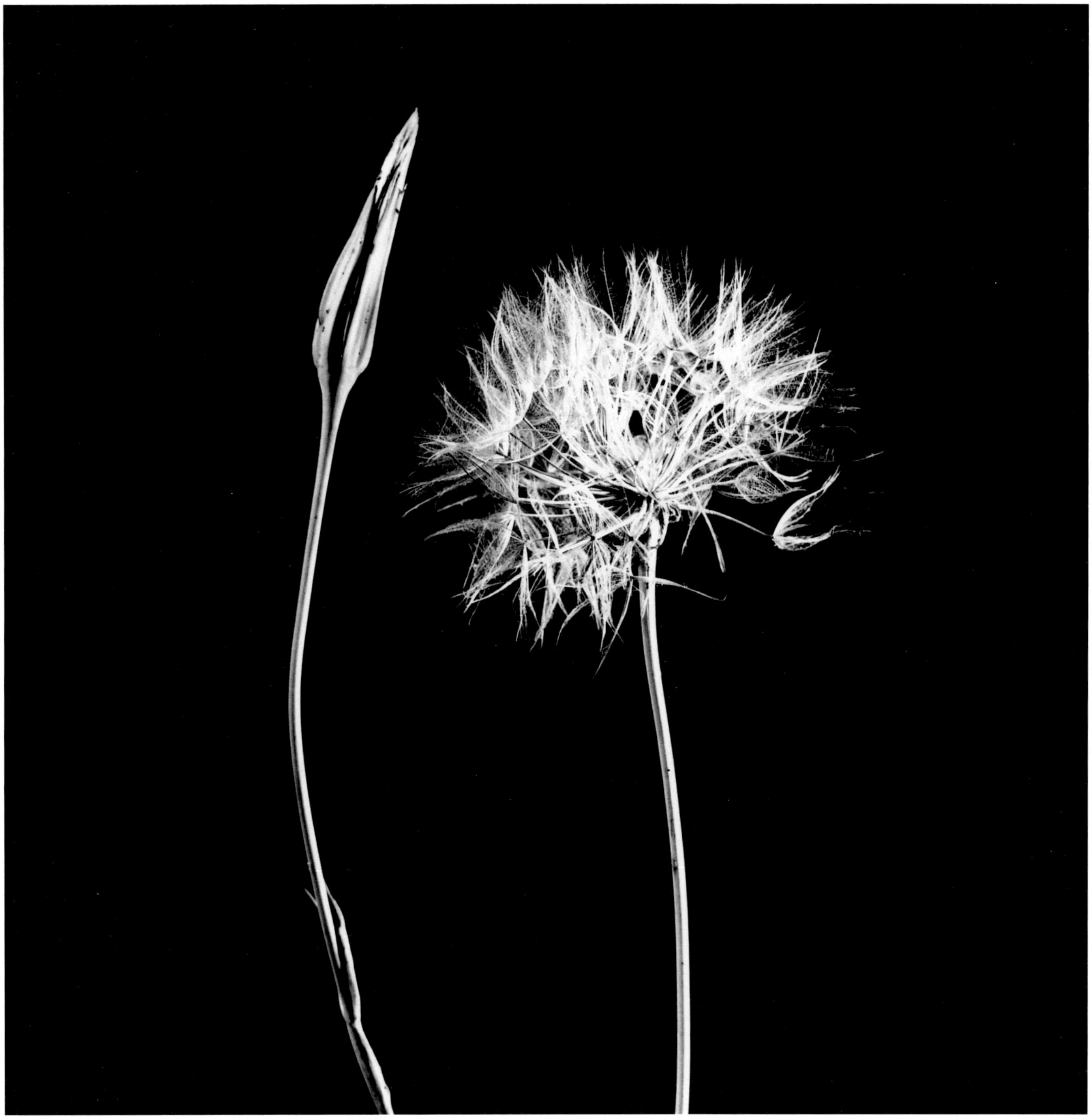

Scorzonera laciniata

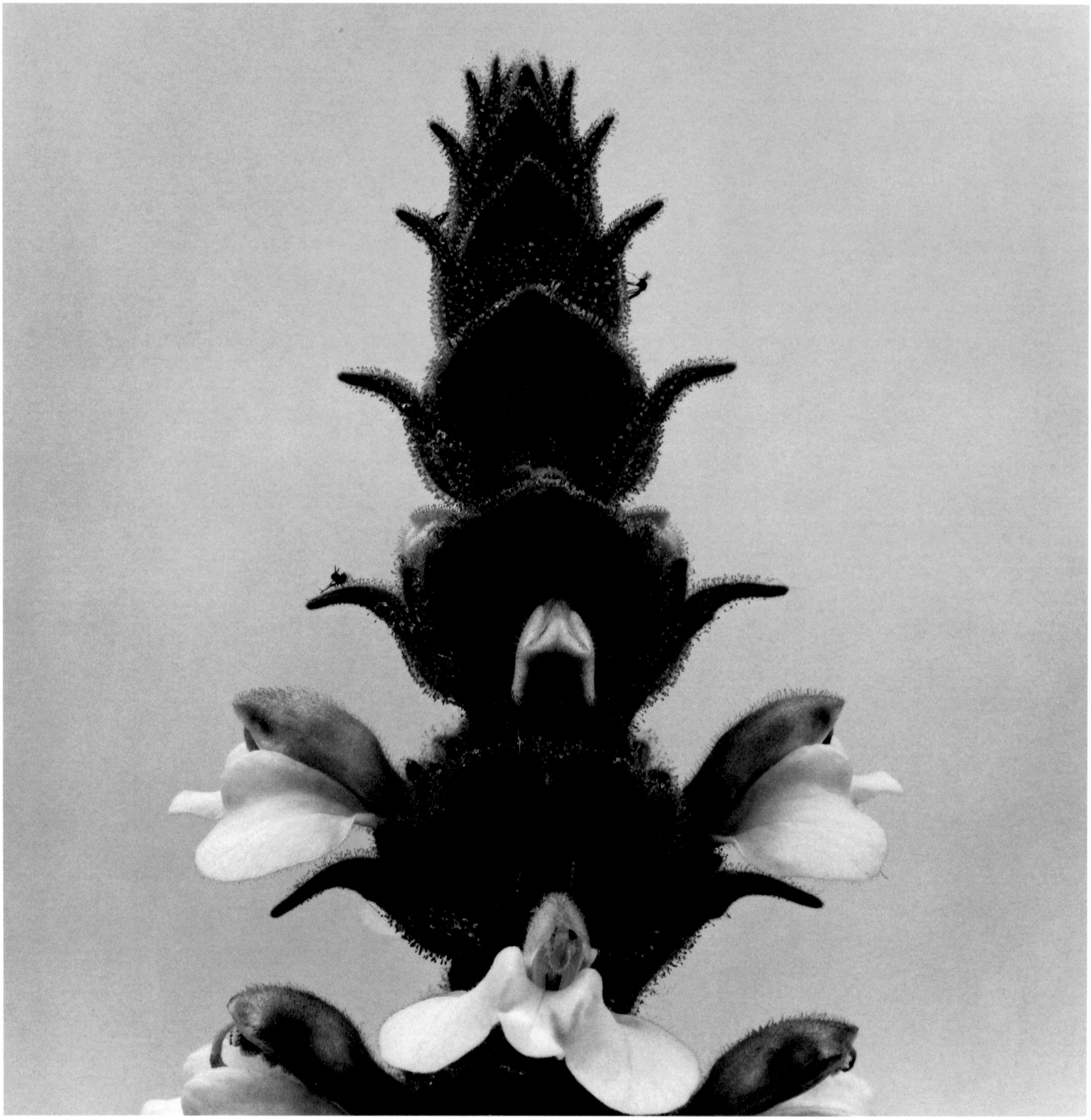

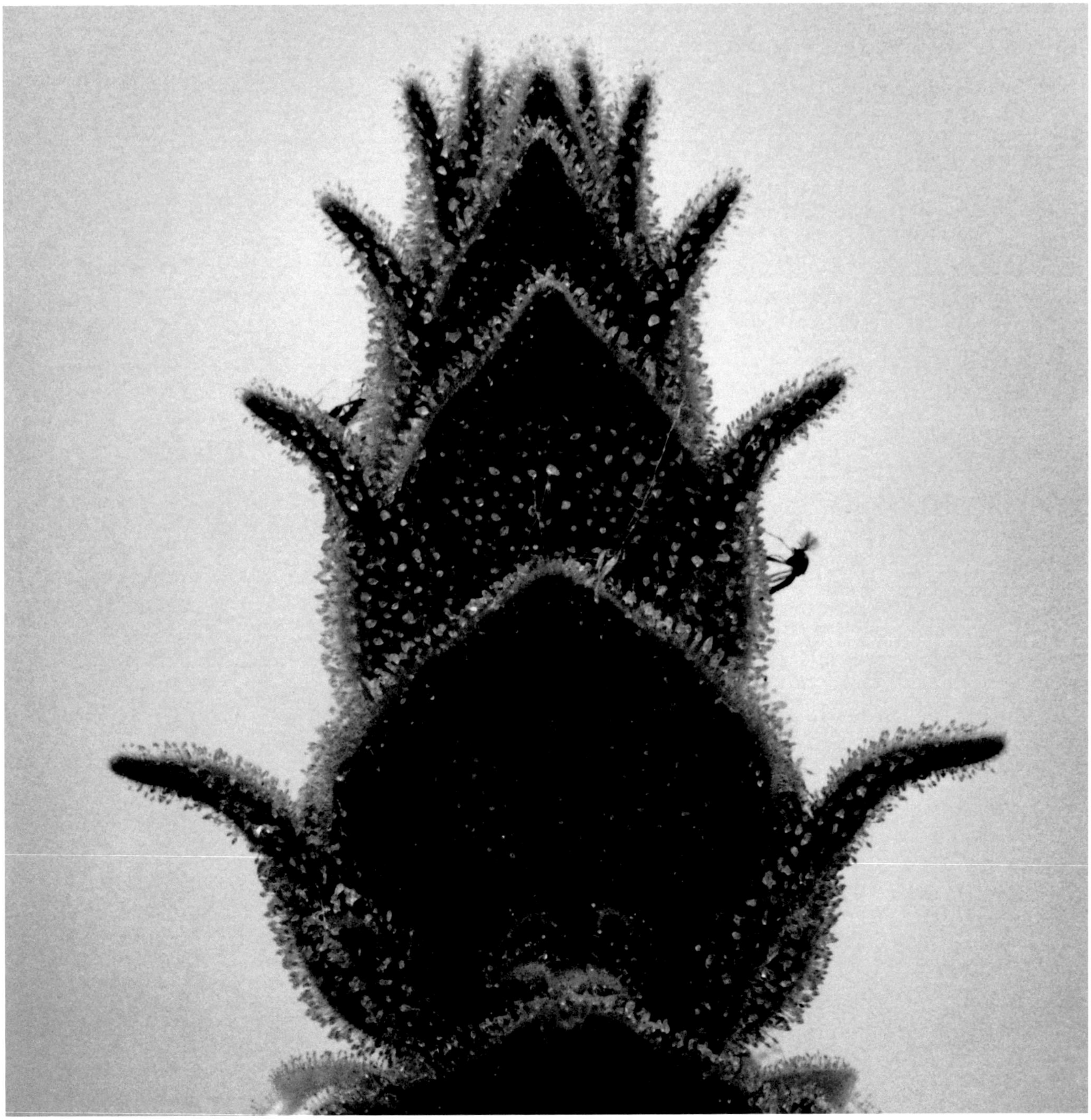

Bellardia trixago

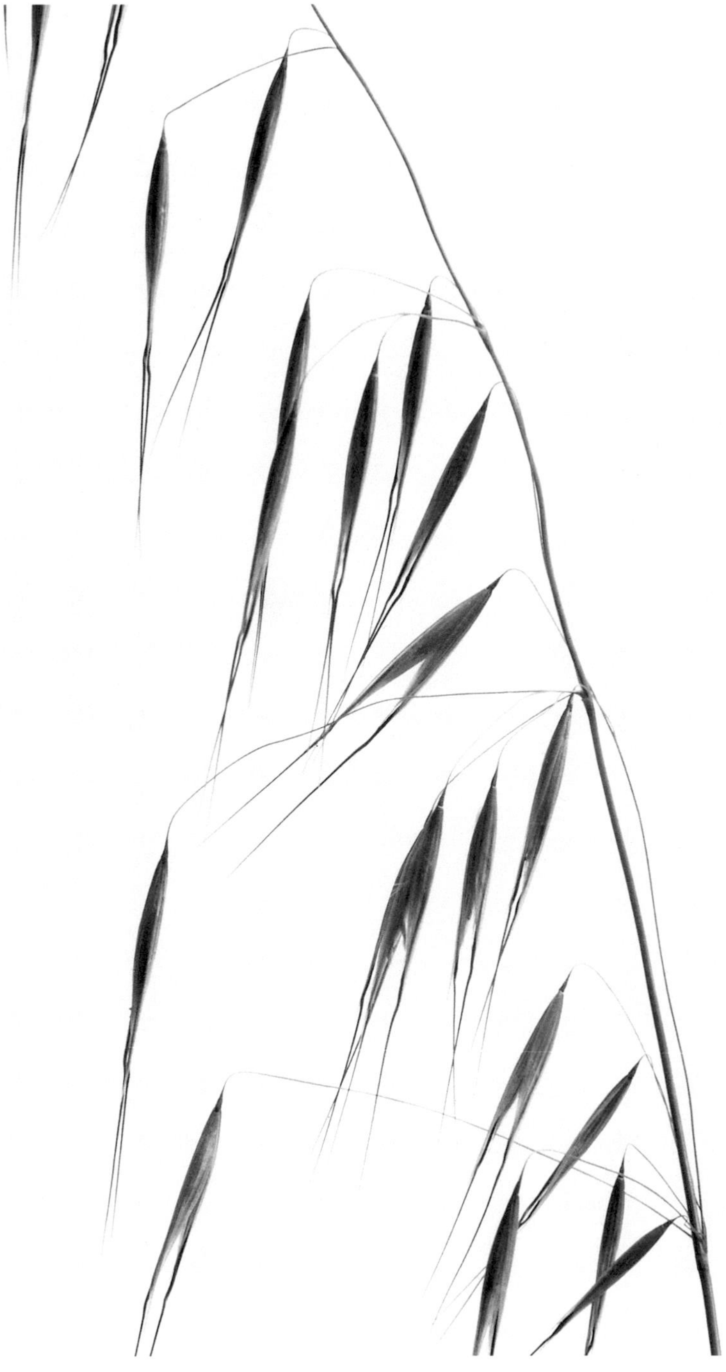

Avena barbata

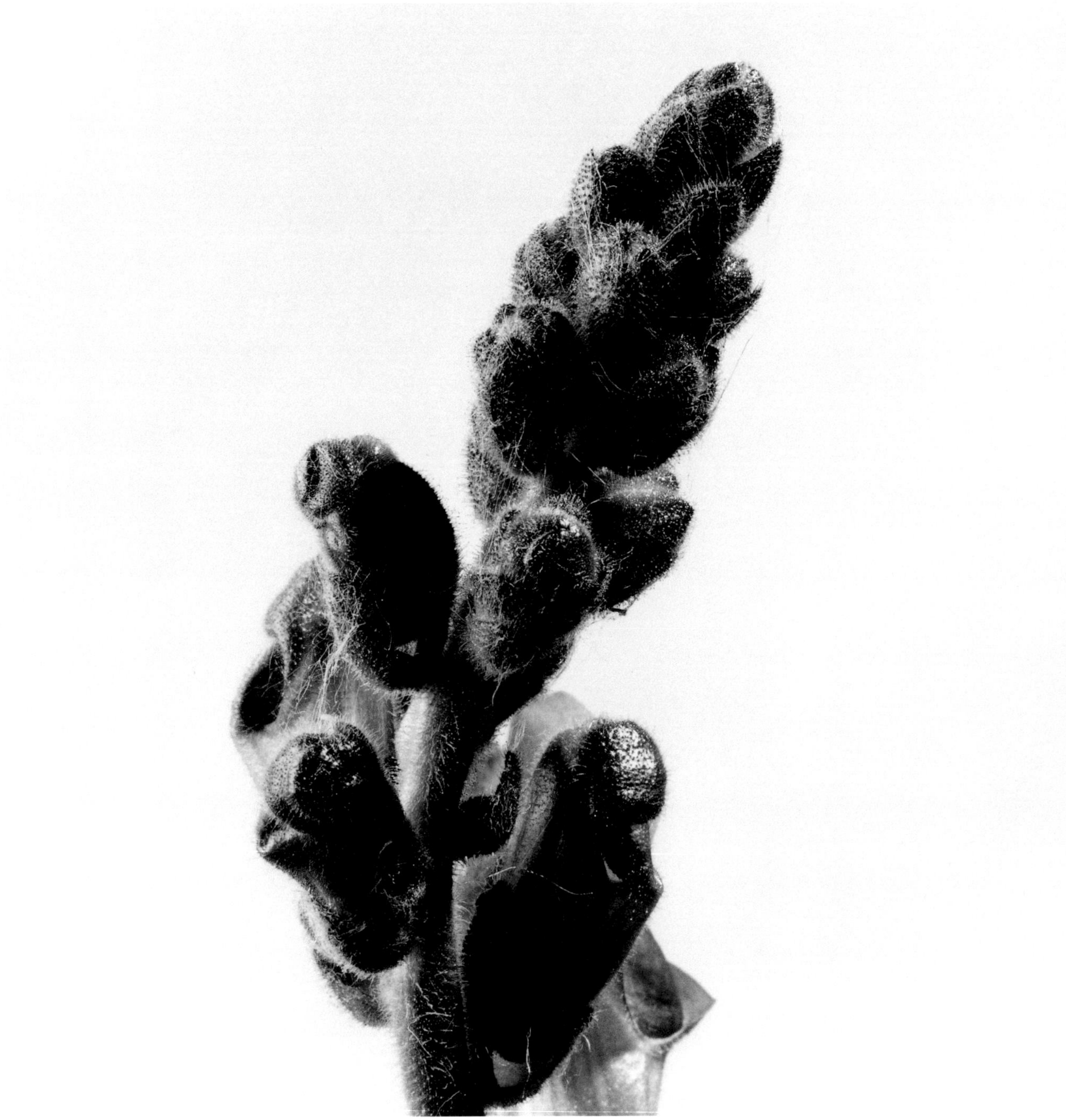

Antirrhinum majus

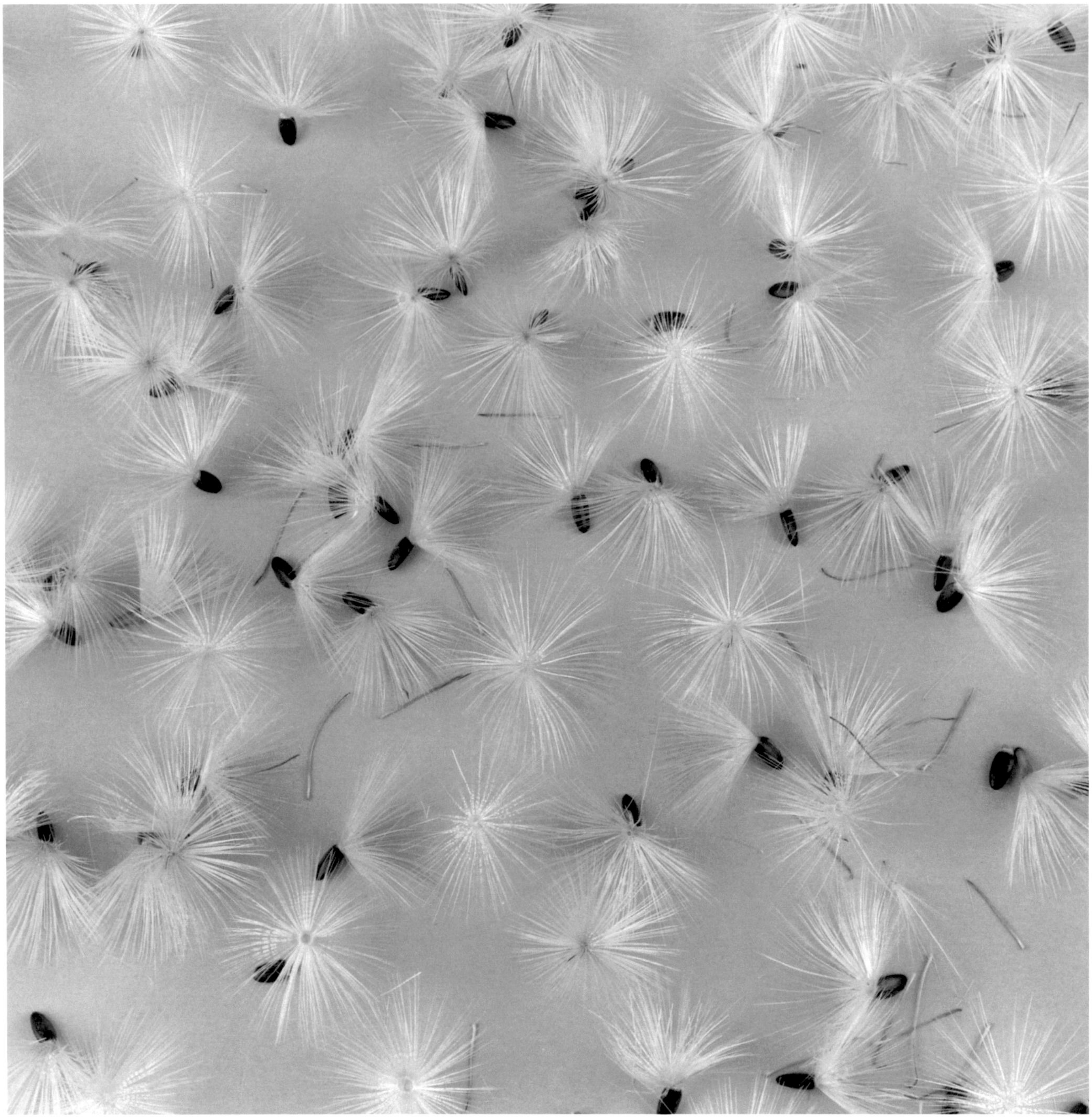

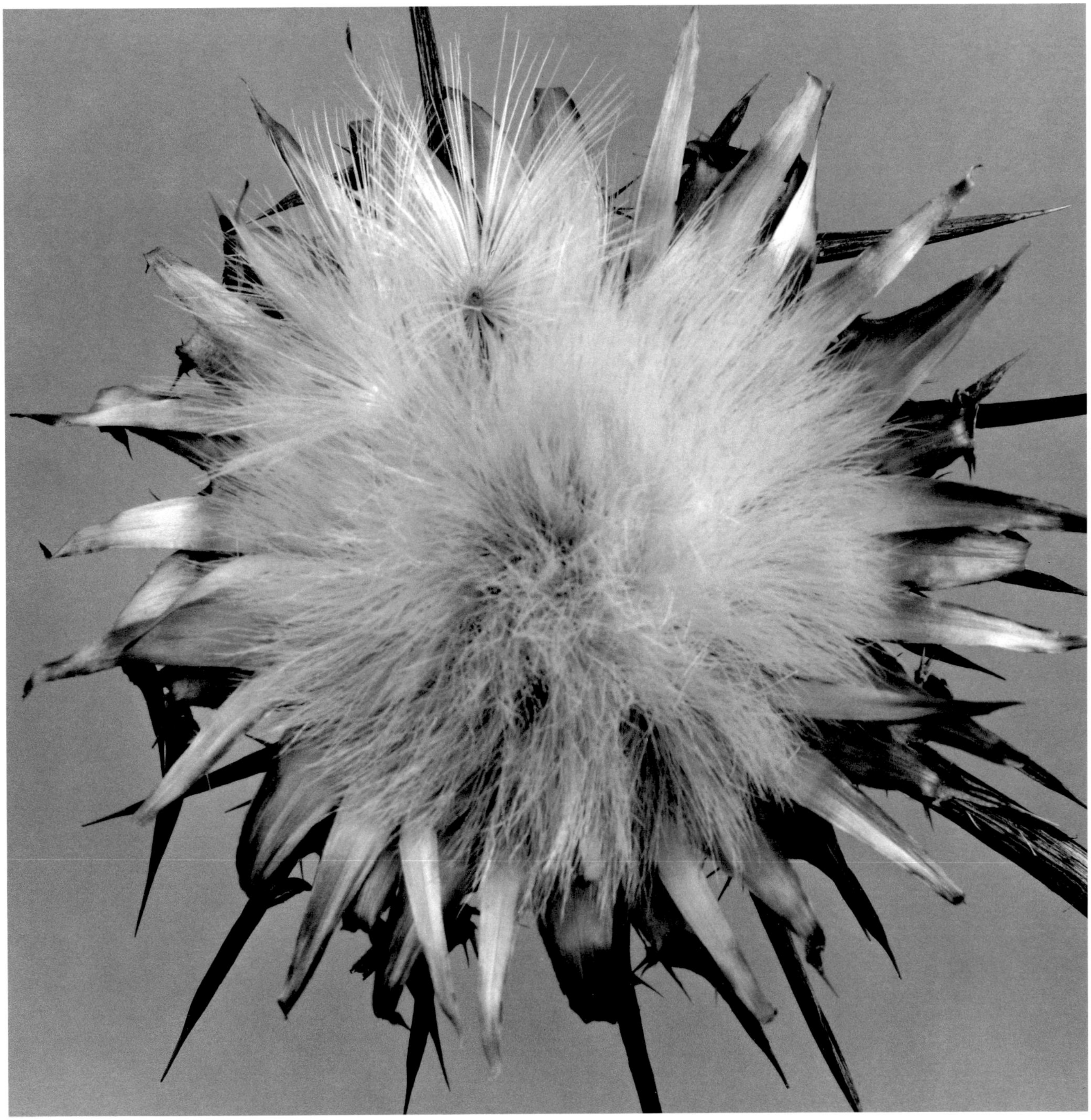

Silybum marianum

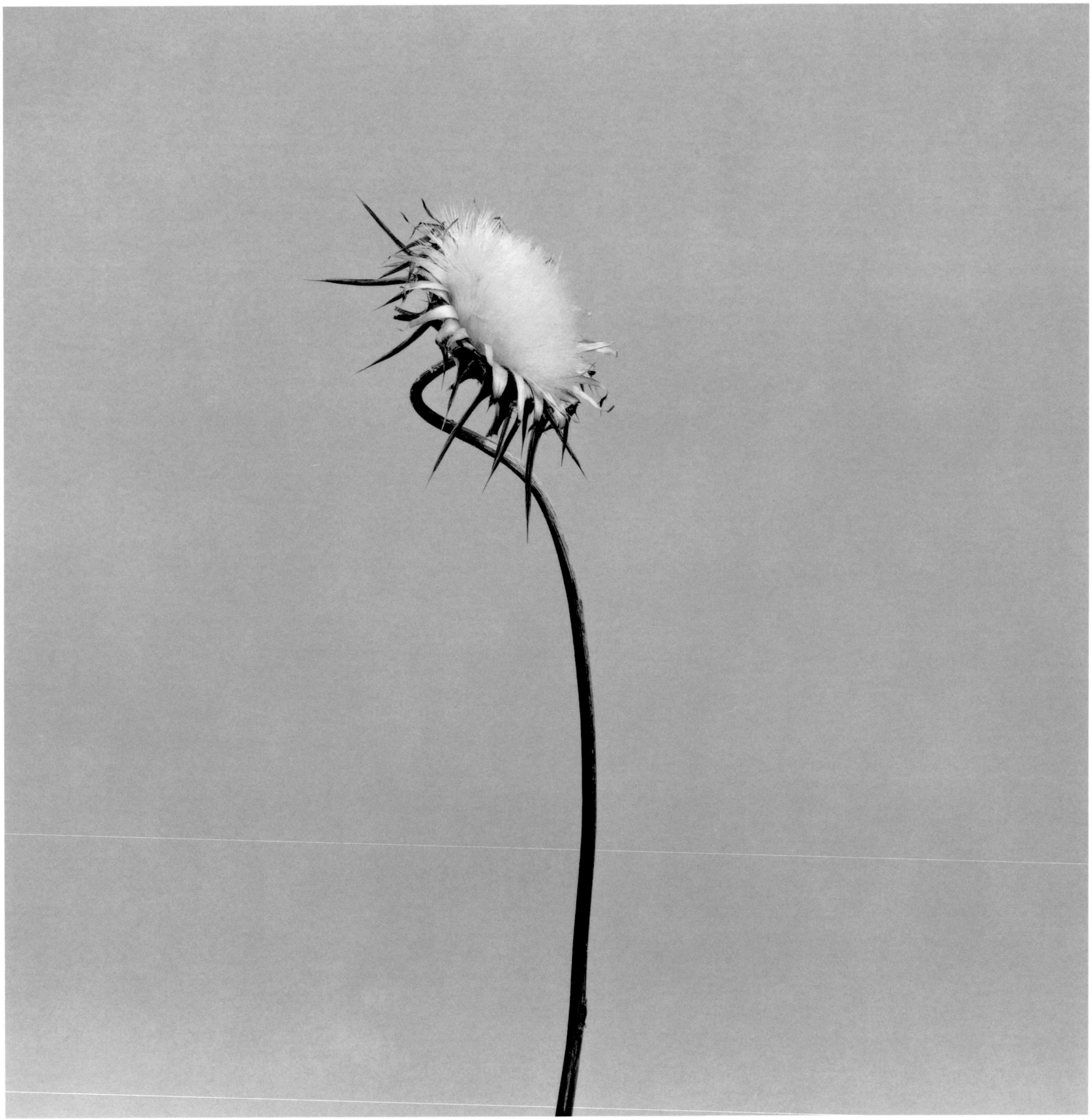

Silybum marianum

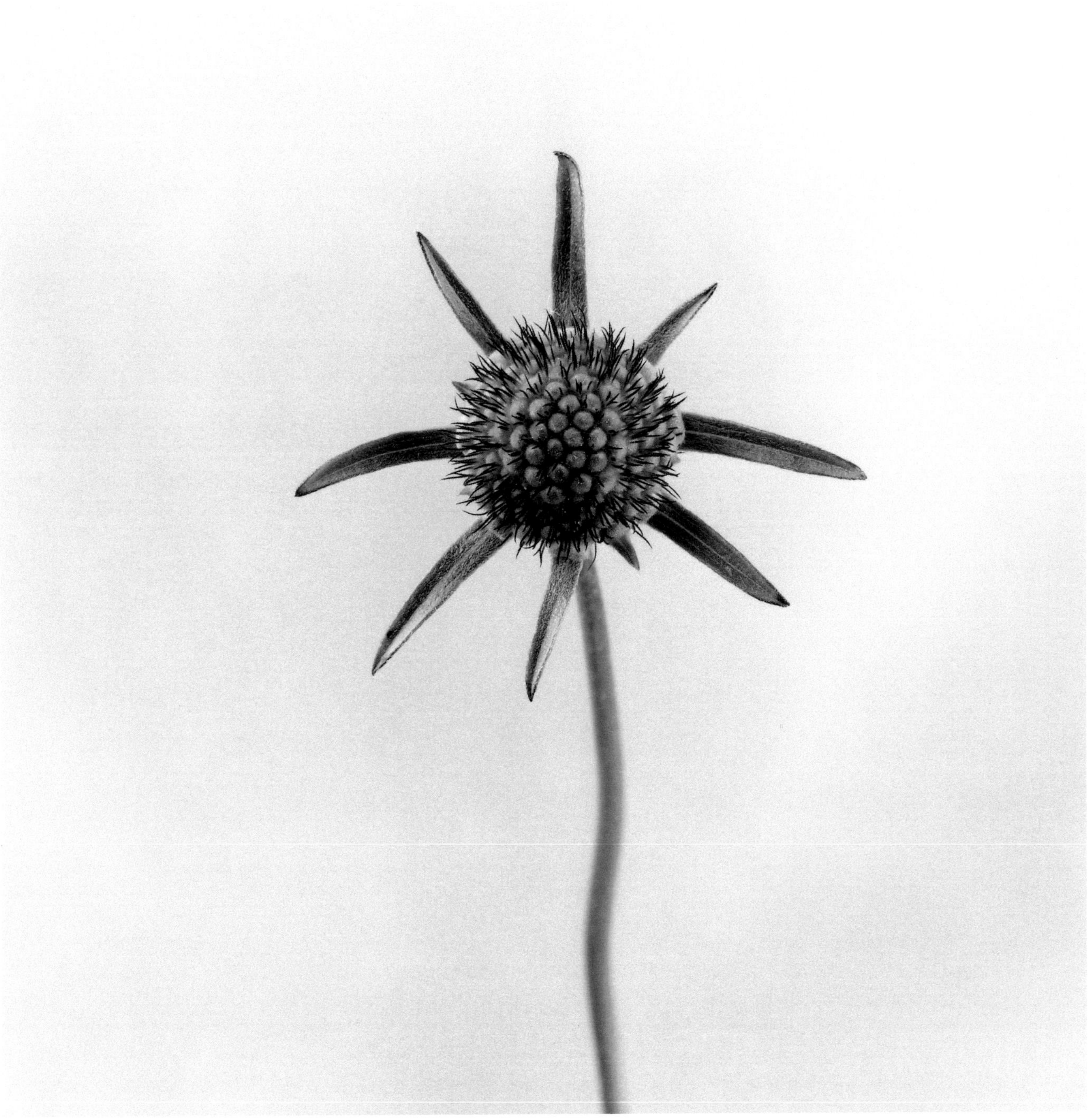

Scabiosa atropurpurea

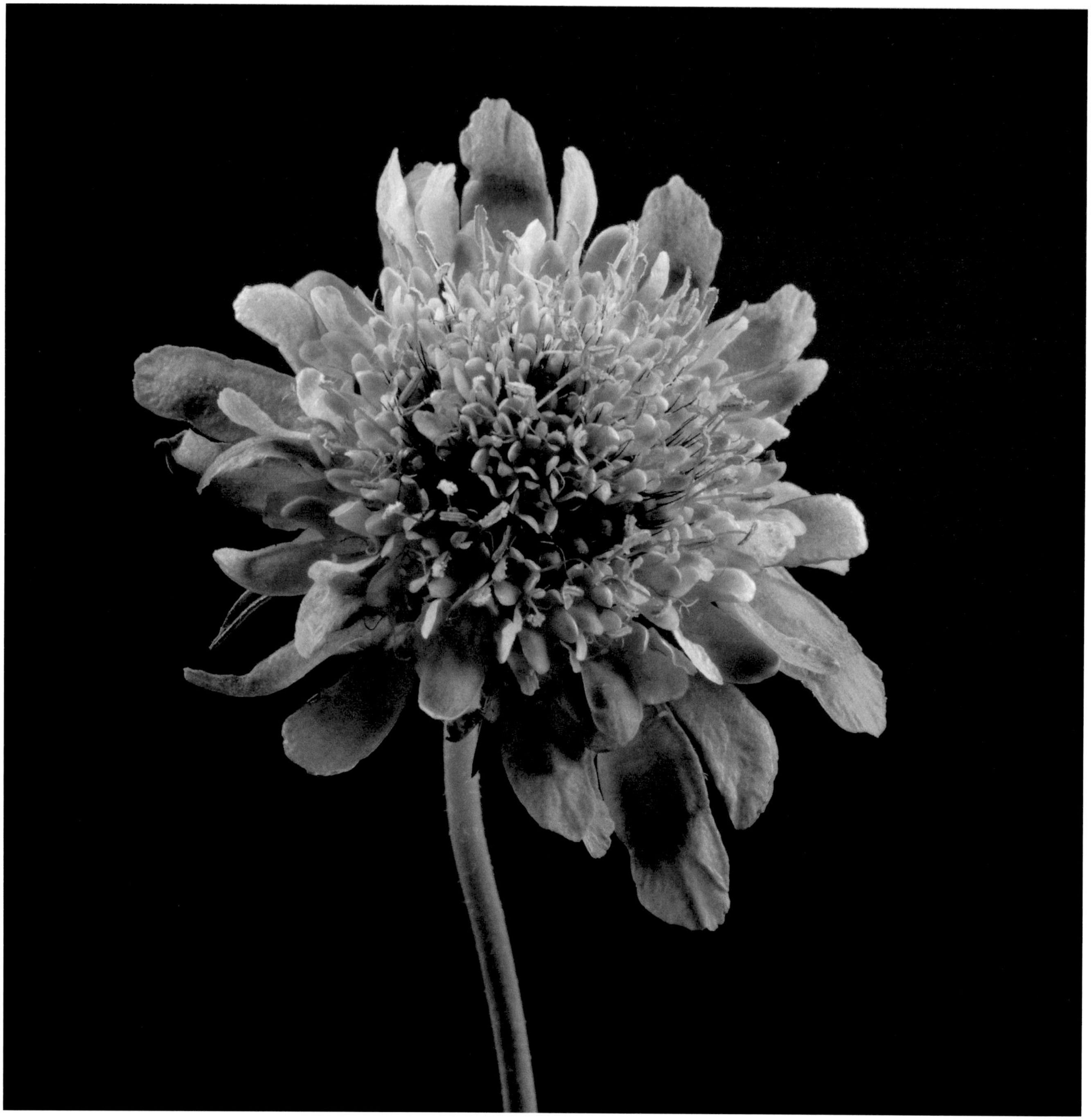

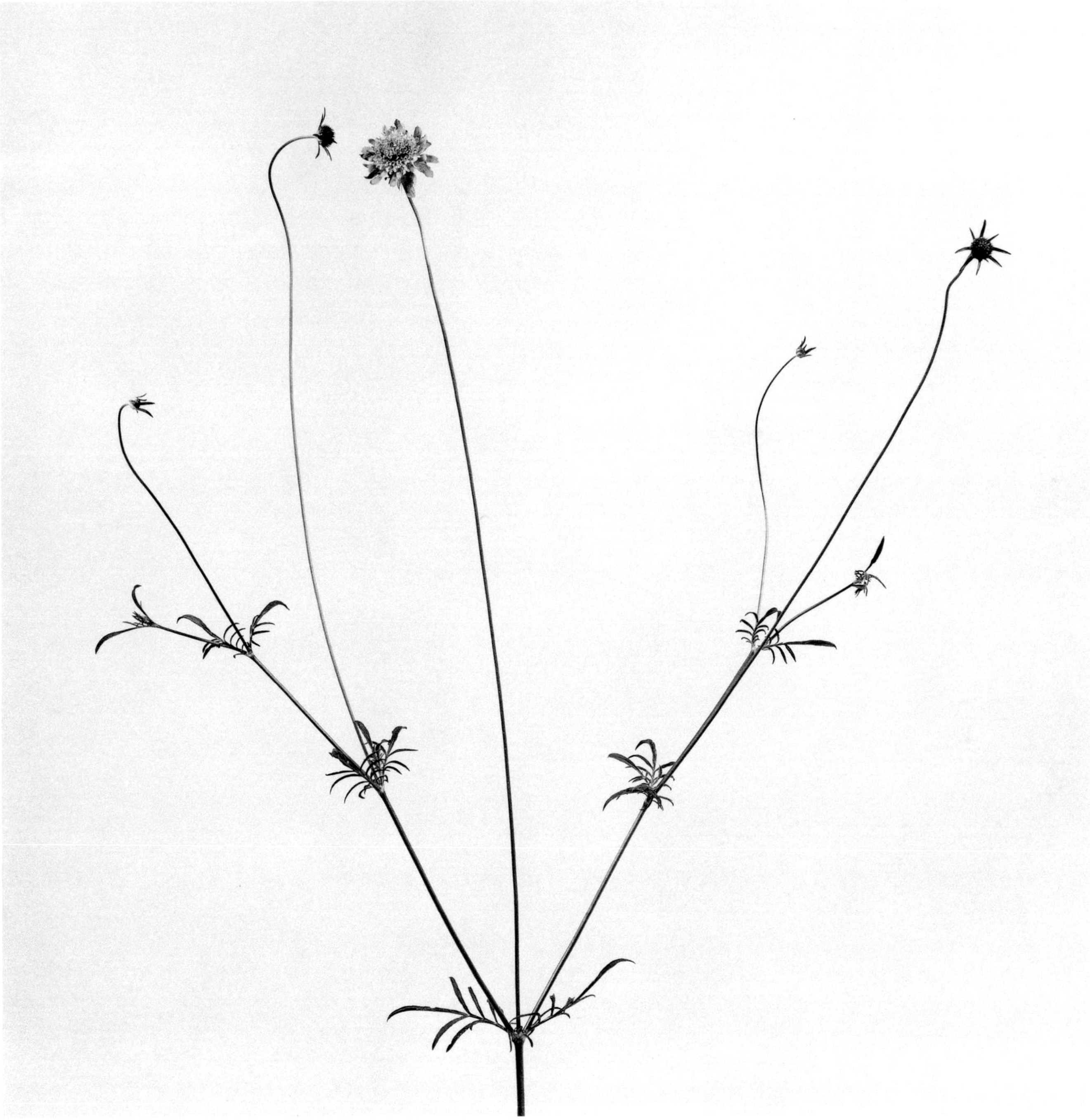

Scabiosa atropurpurea

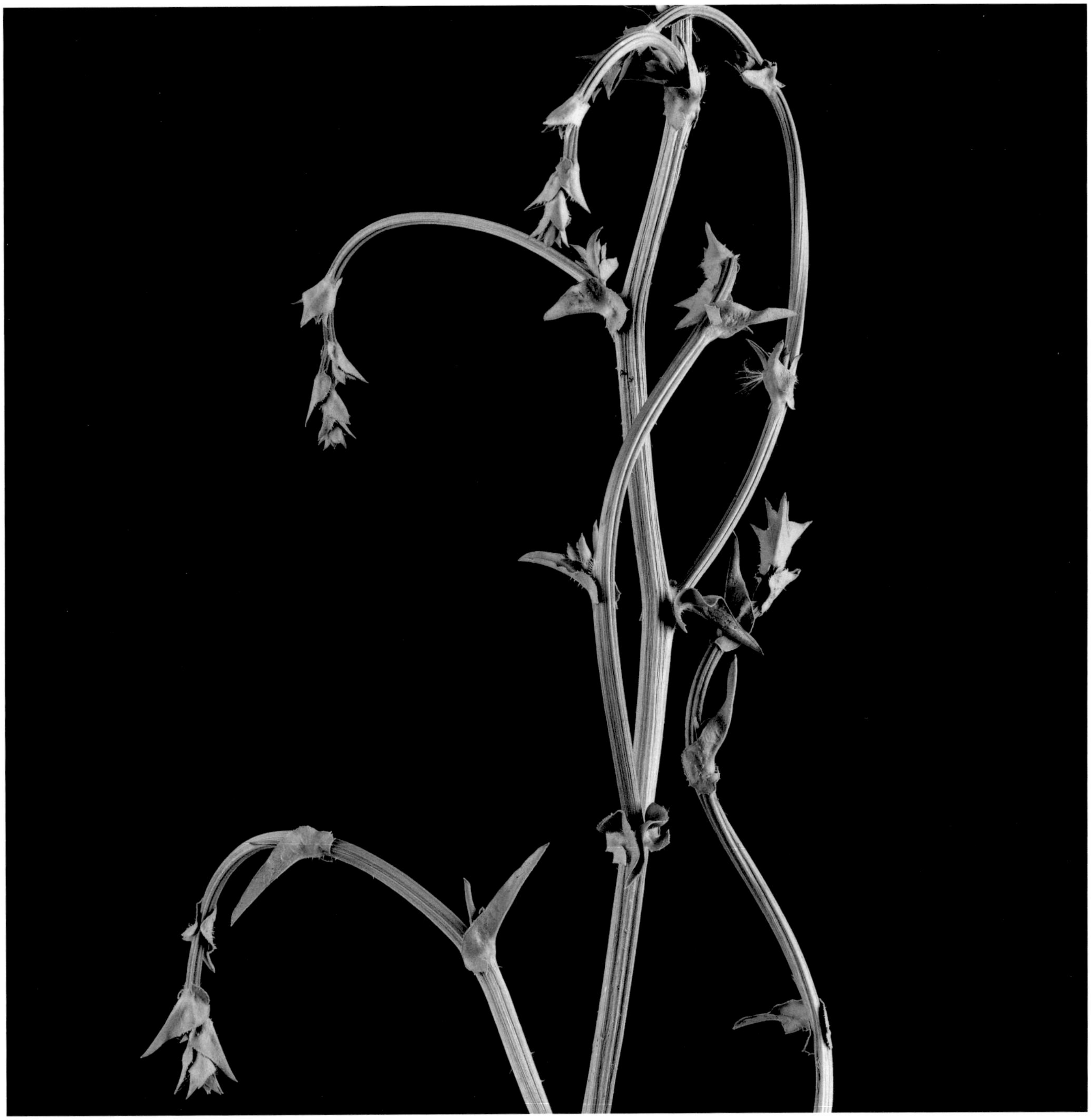

Sp. ign.

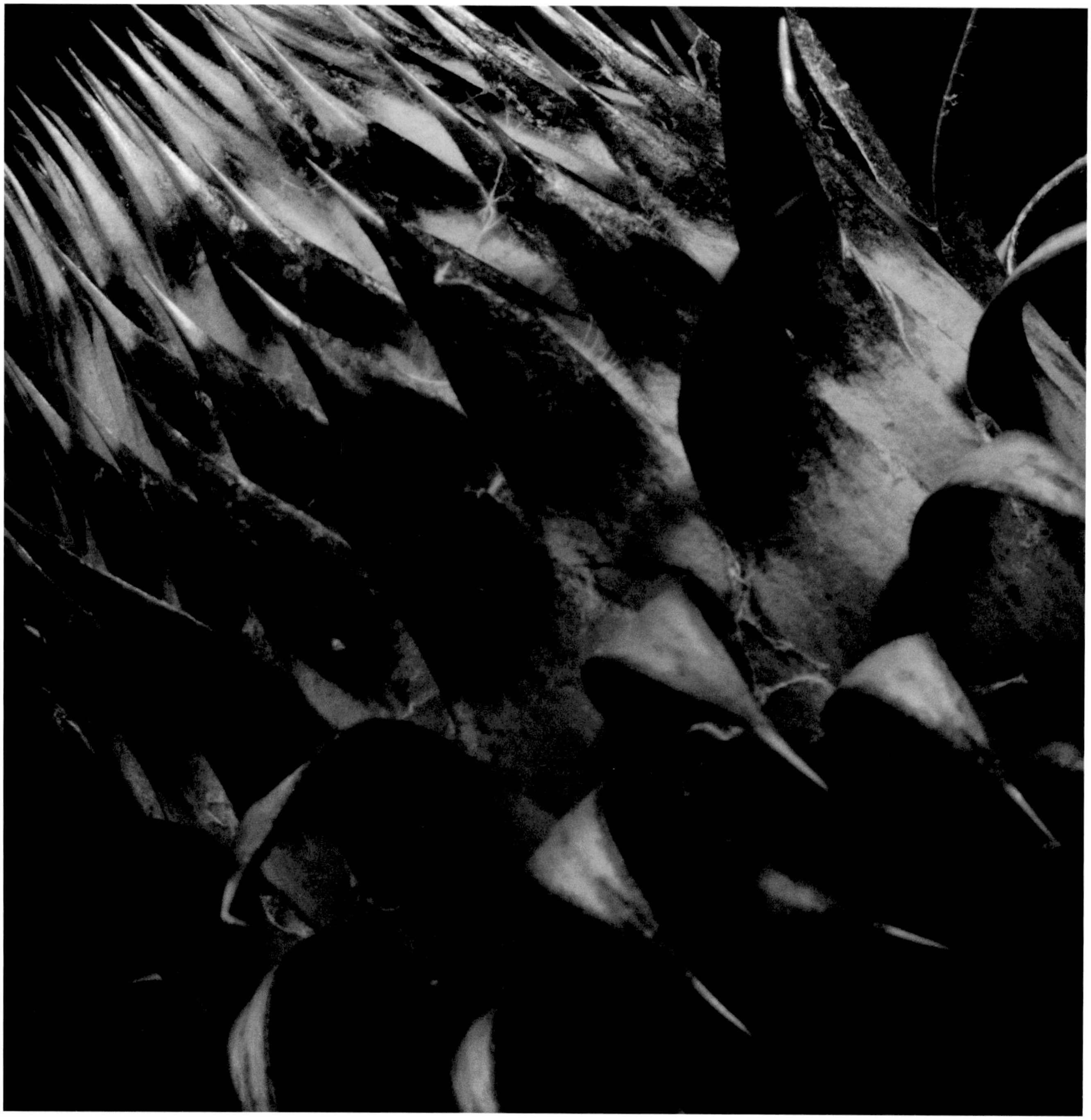

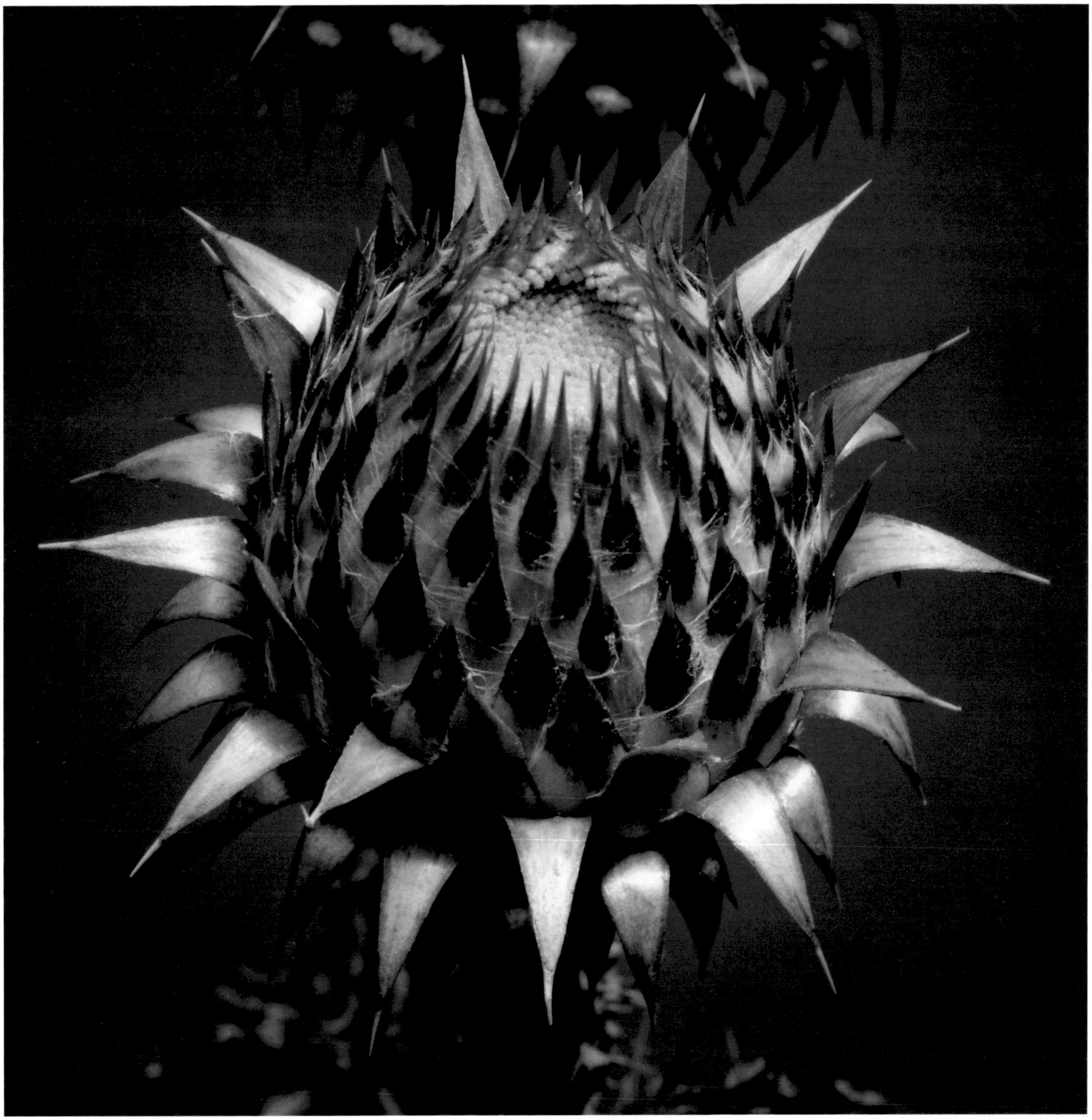

Cynara cardunculus

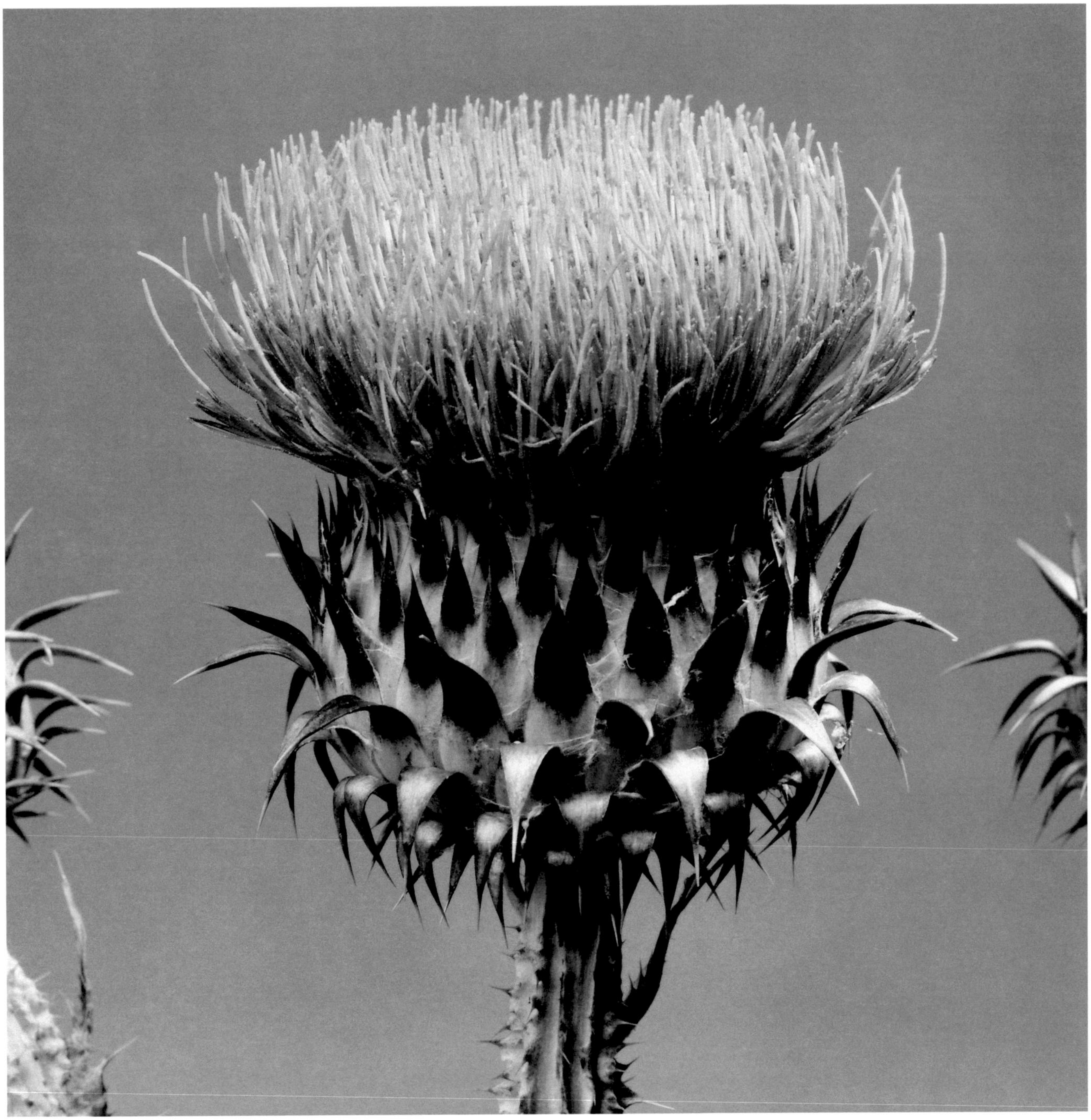

Cynara cardunculus

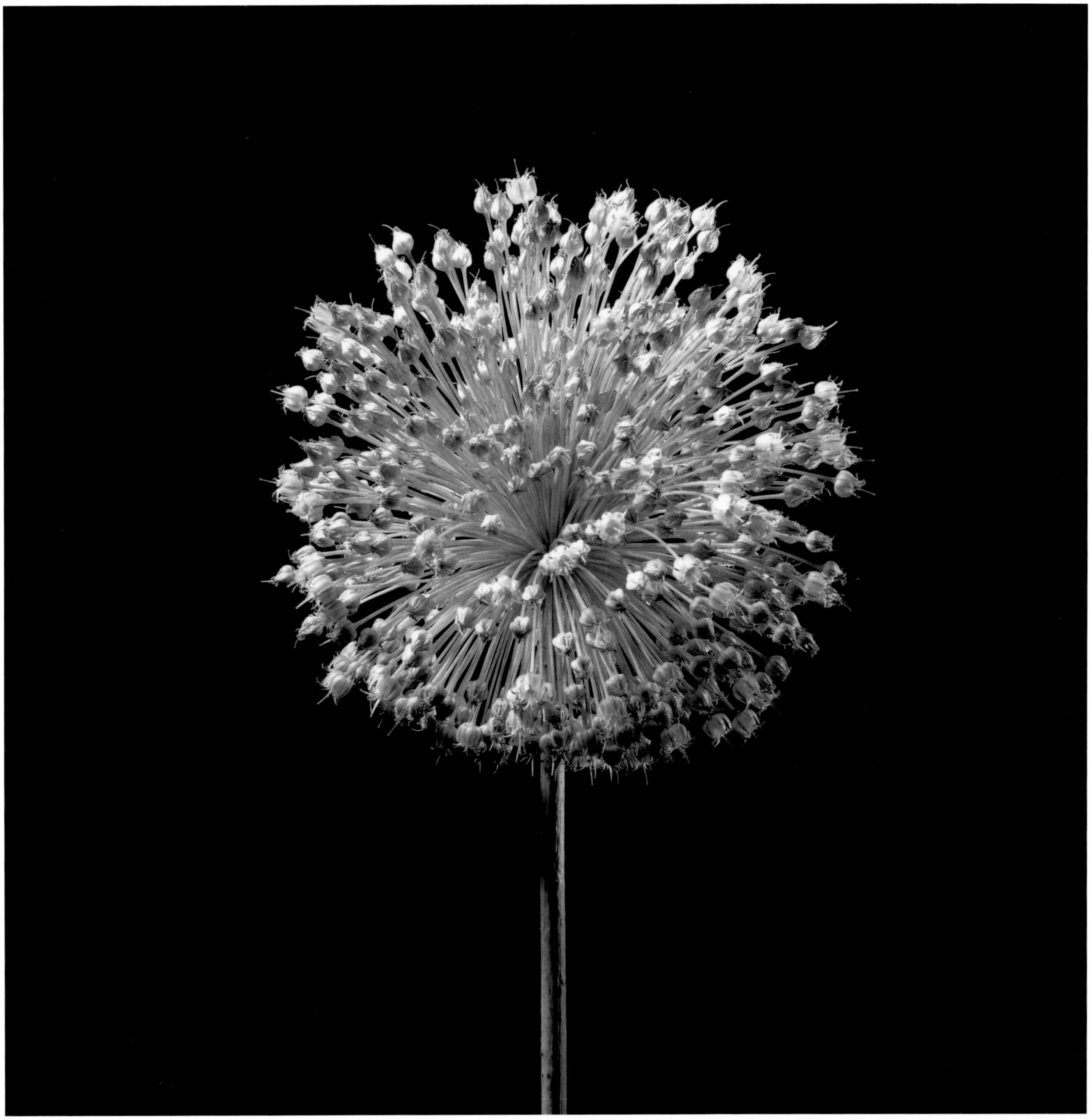

Allium ampeloprasum

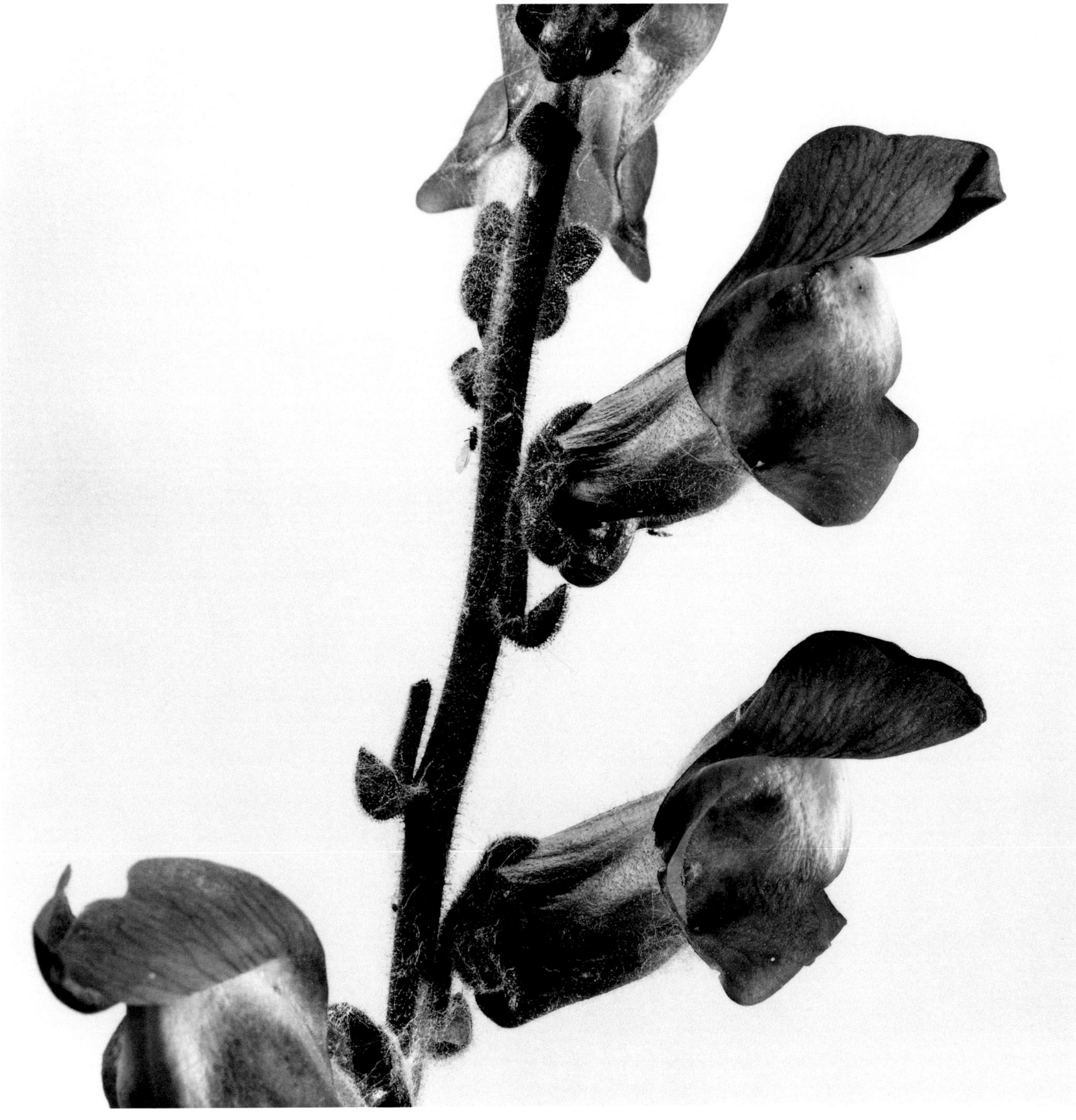

Antirrhinum majus

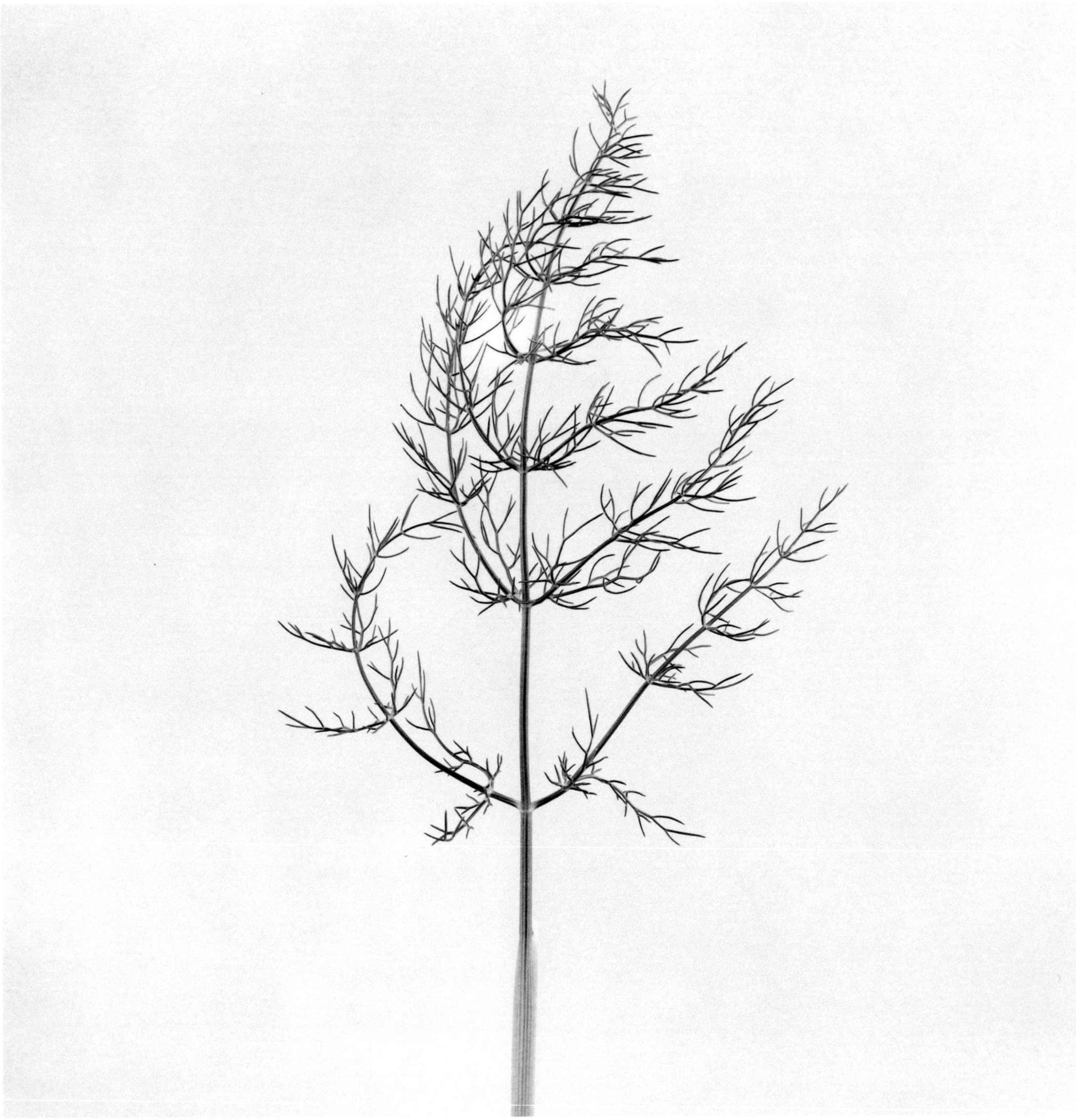

Foeniculum vulgare

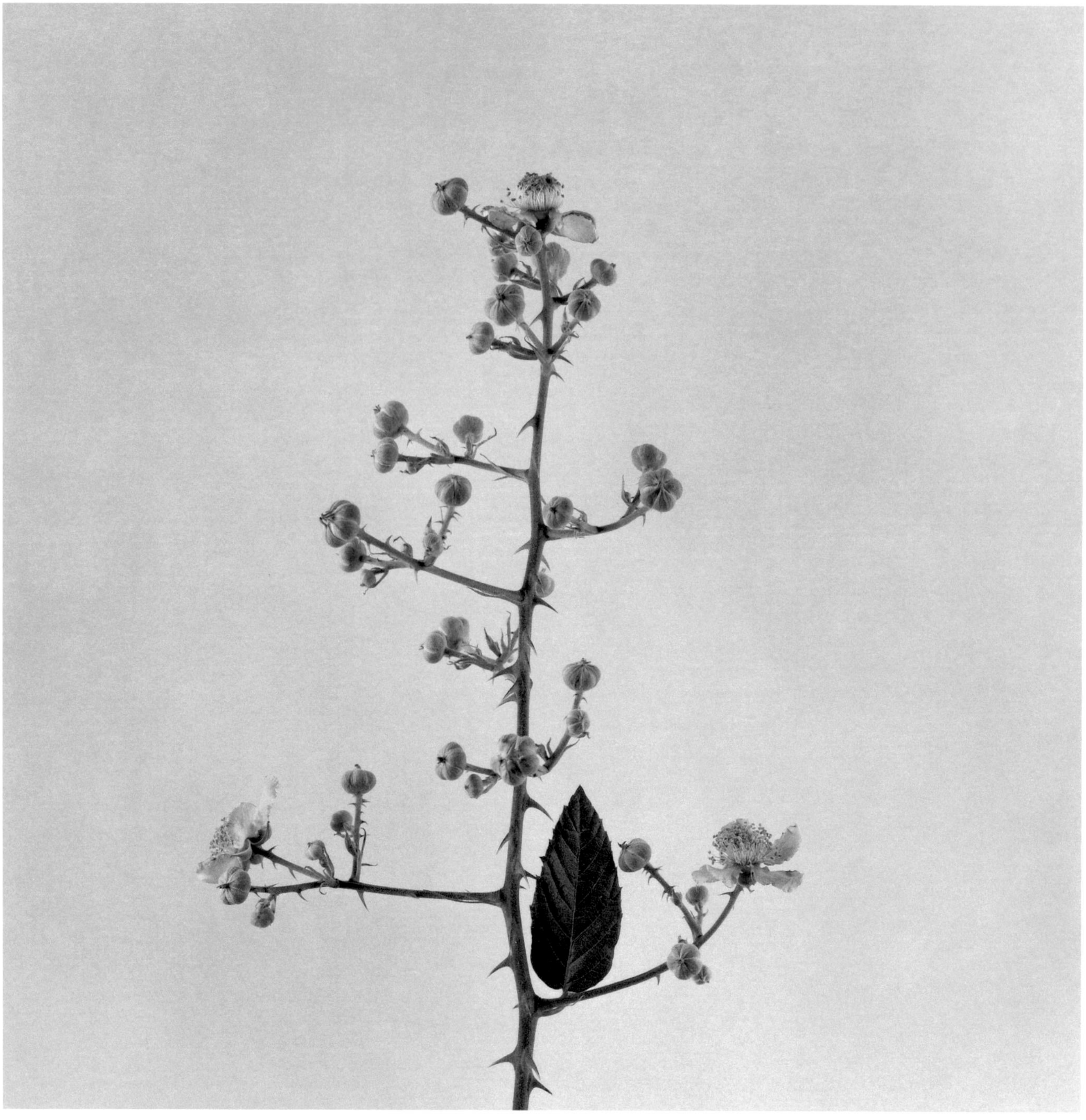

Rubus ulmifolius

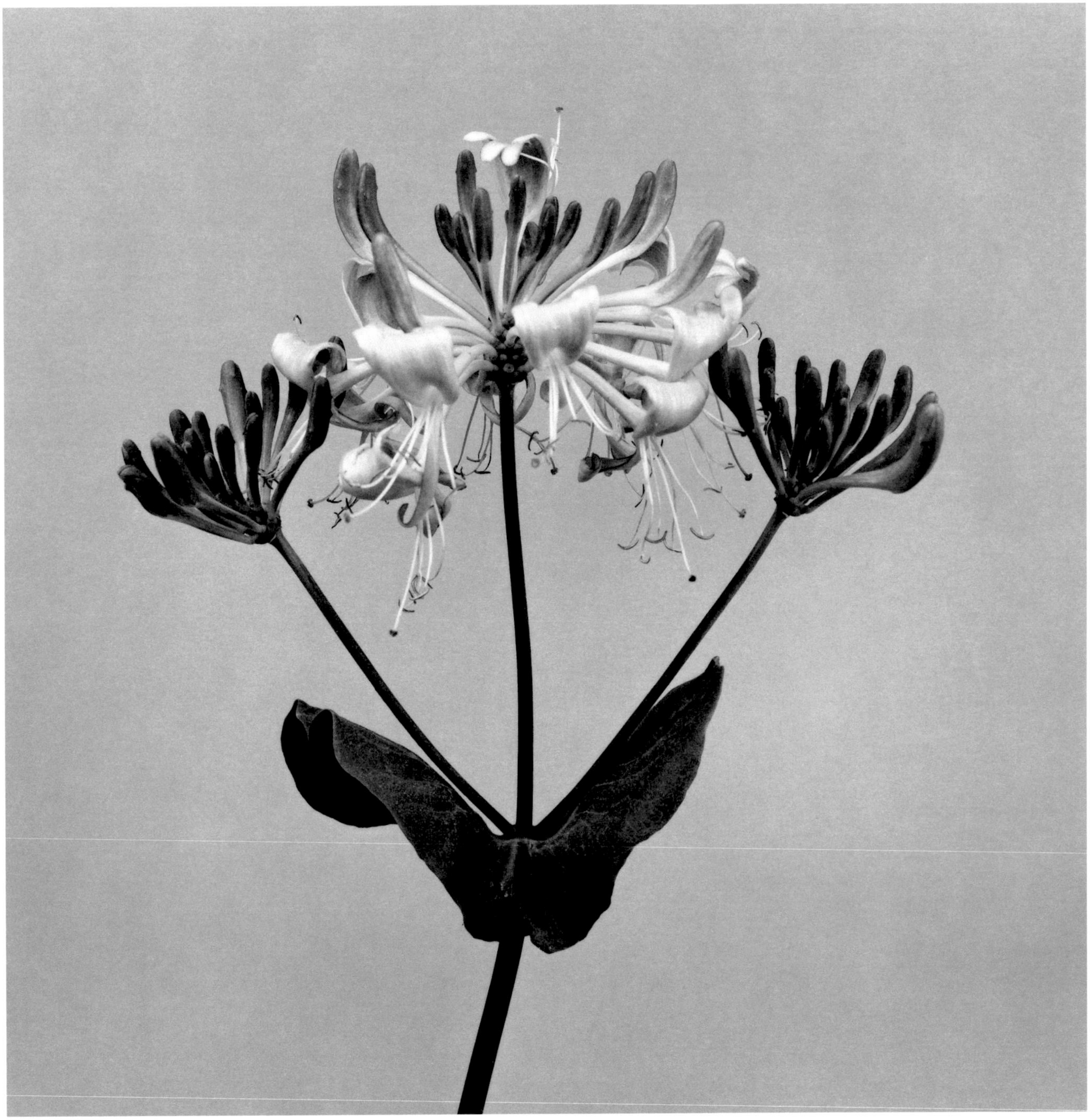

Lonicera implexa

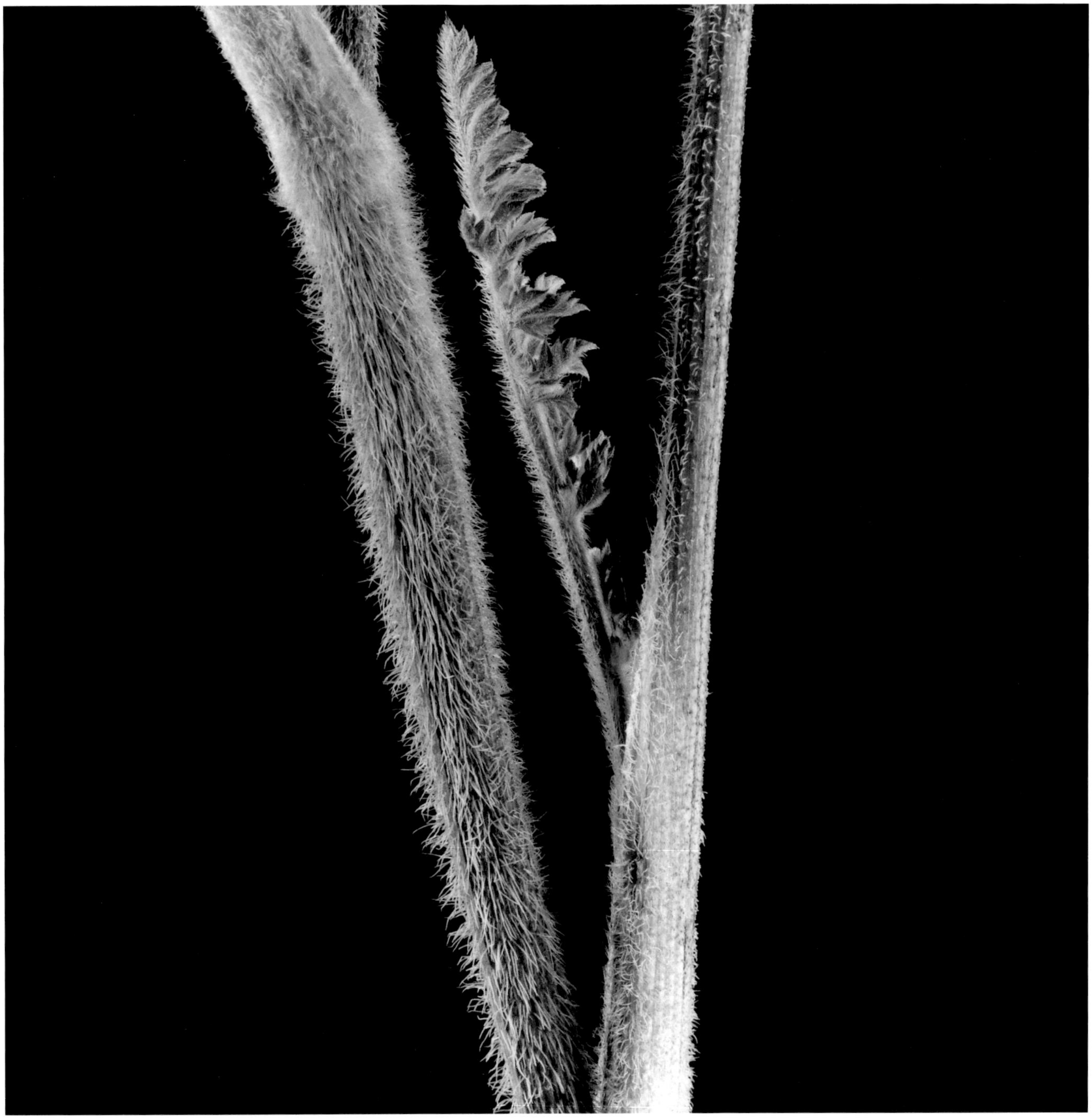

Daucus carota

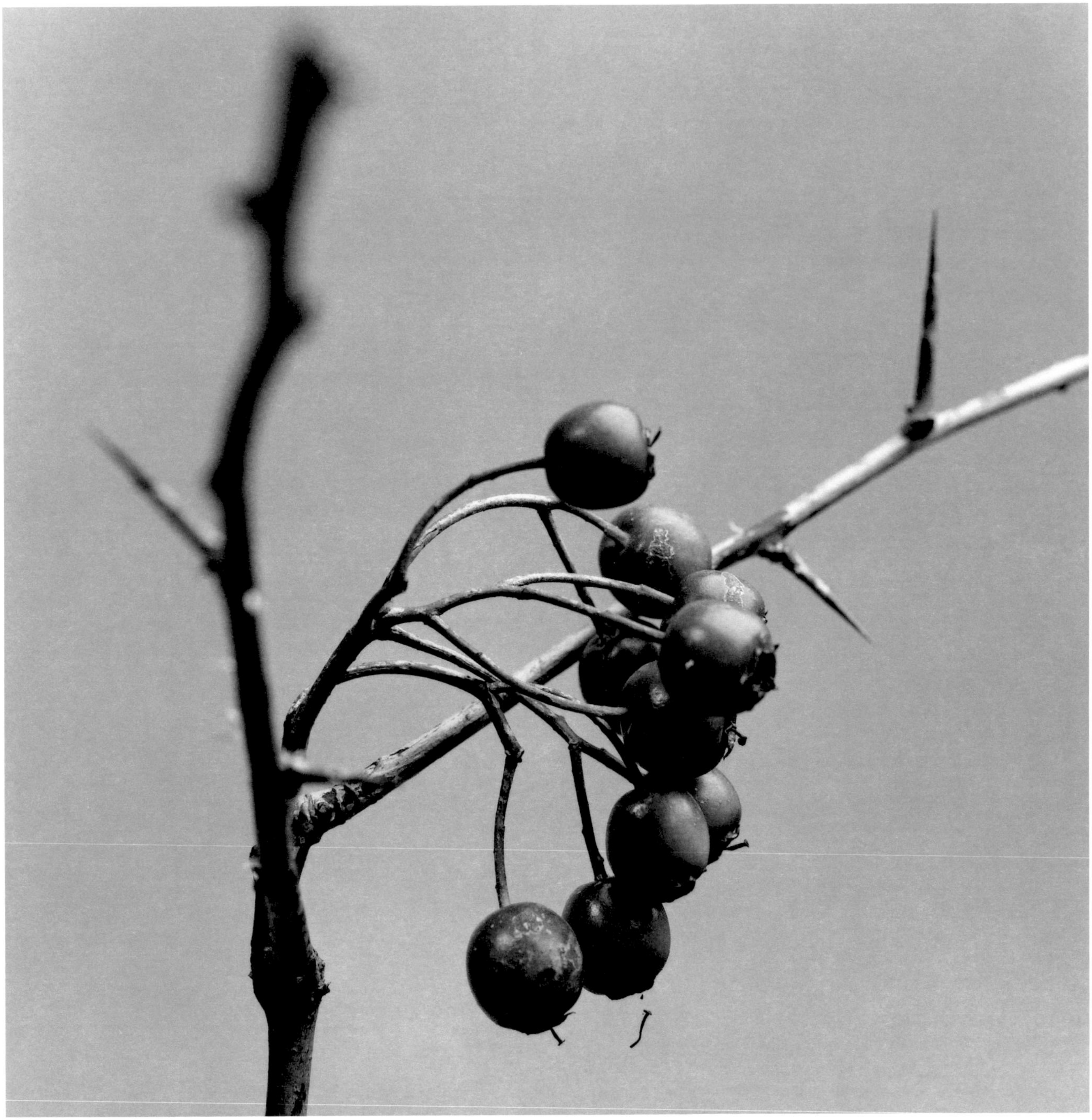

Crataegus monogyna

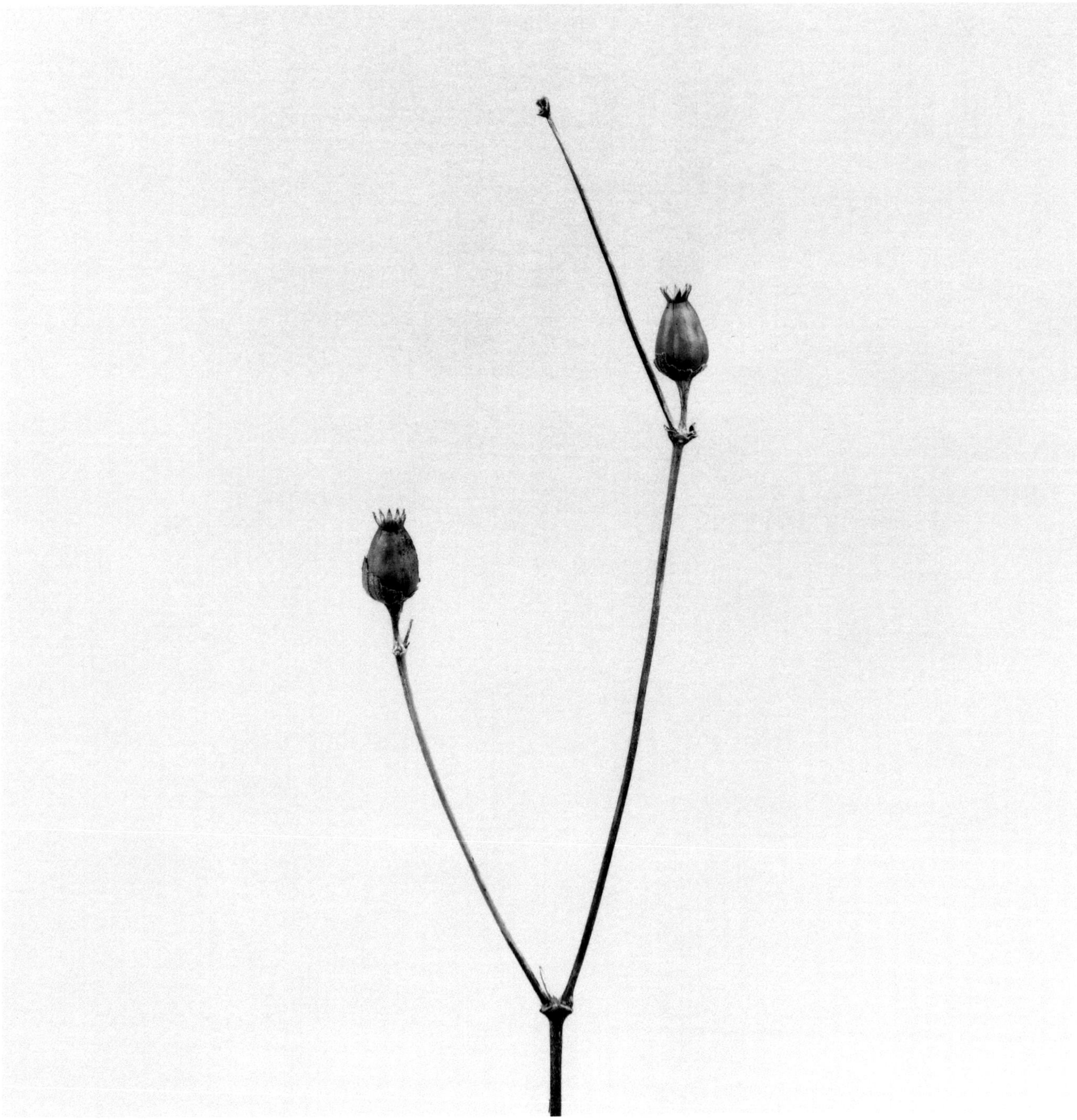

Silene latifolia

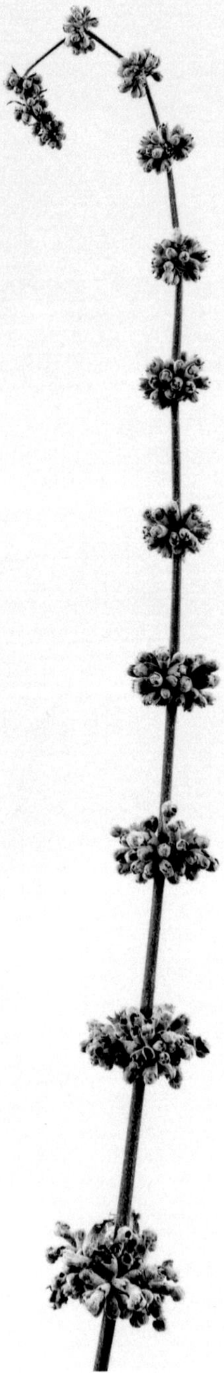

Vitex agnus-castus

Seseli tortuosum

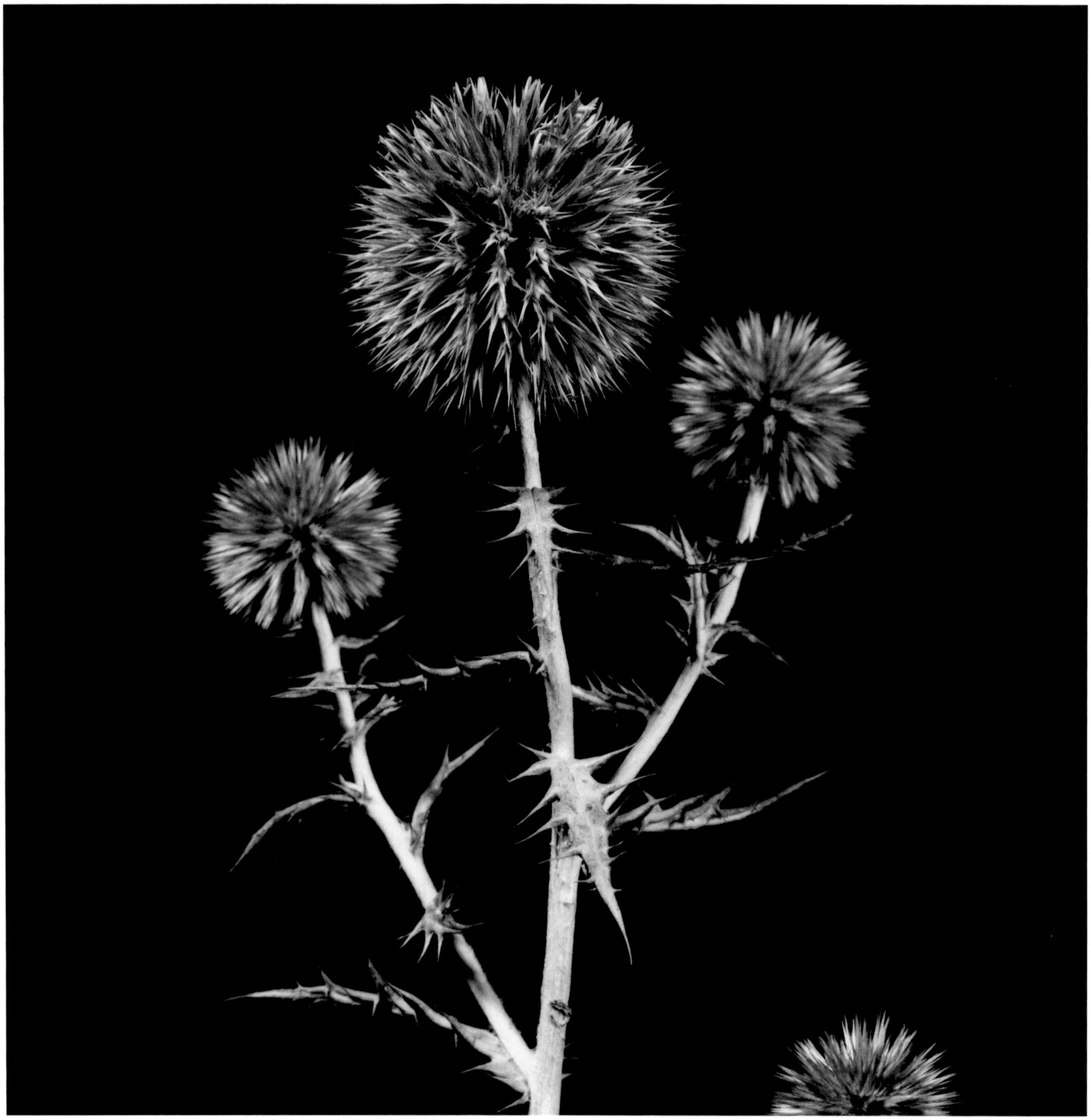

Echinops ritro

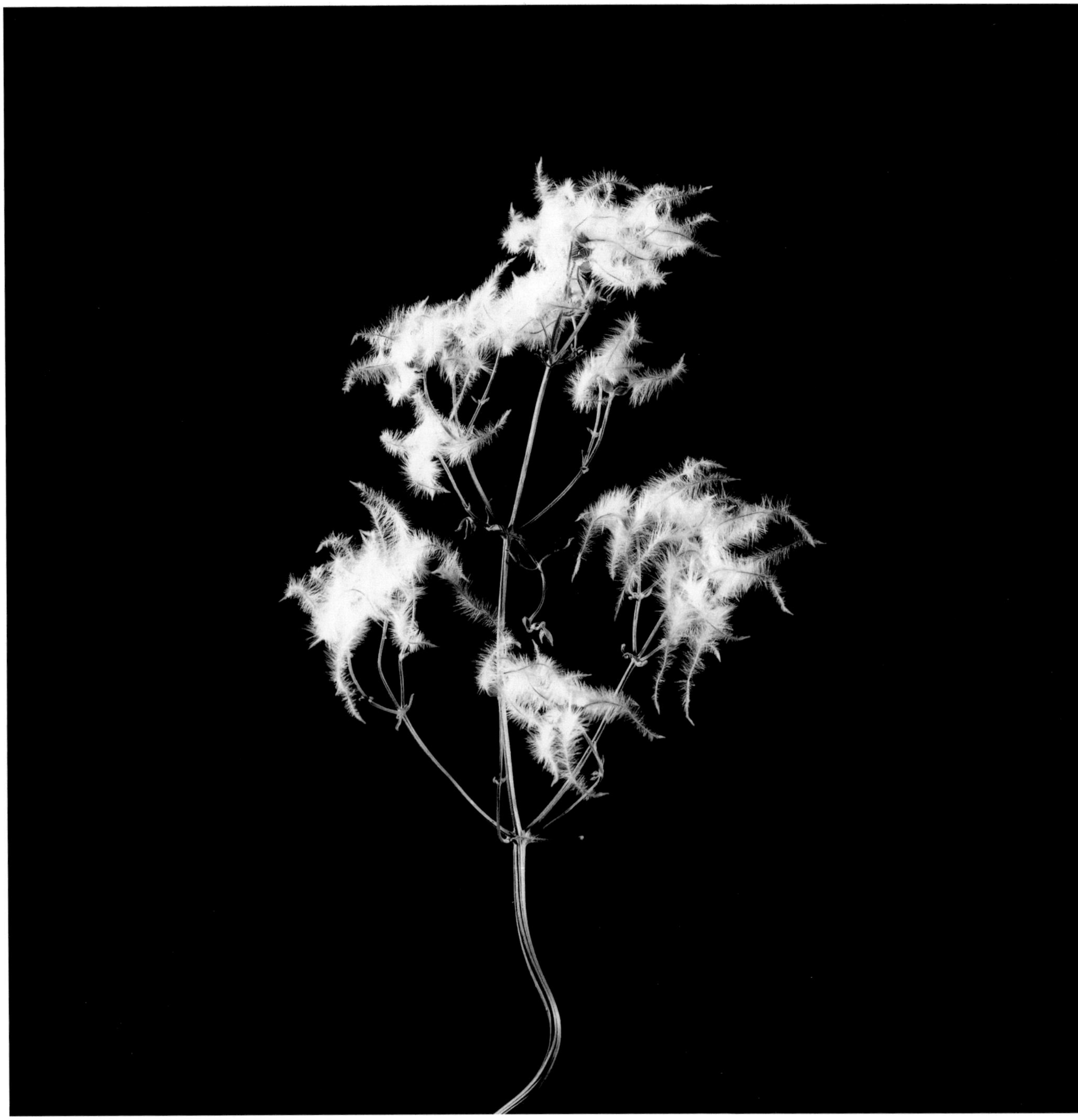

Clematis flammula

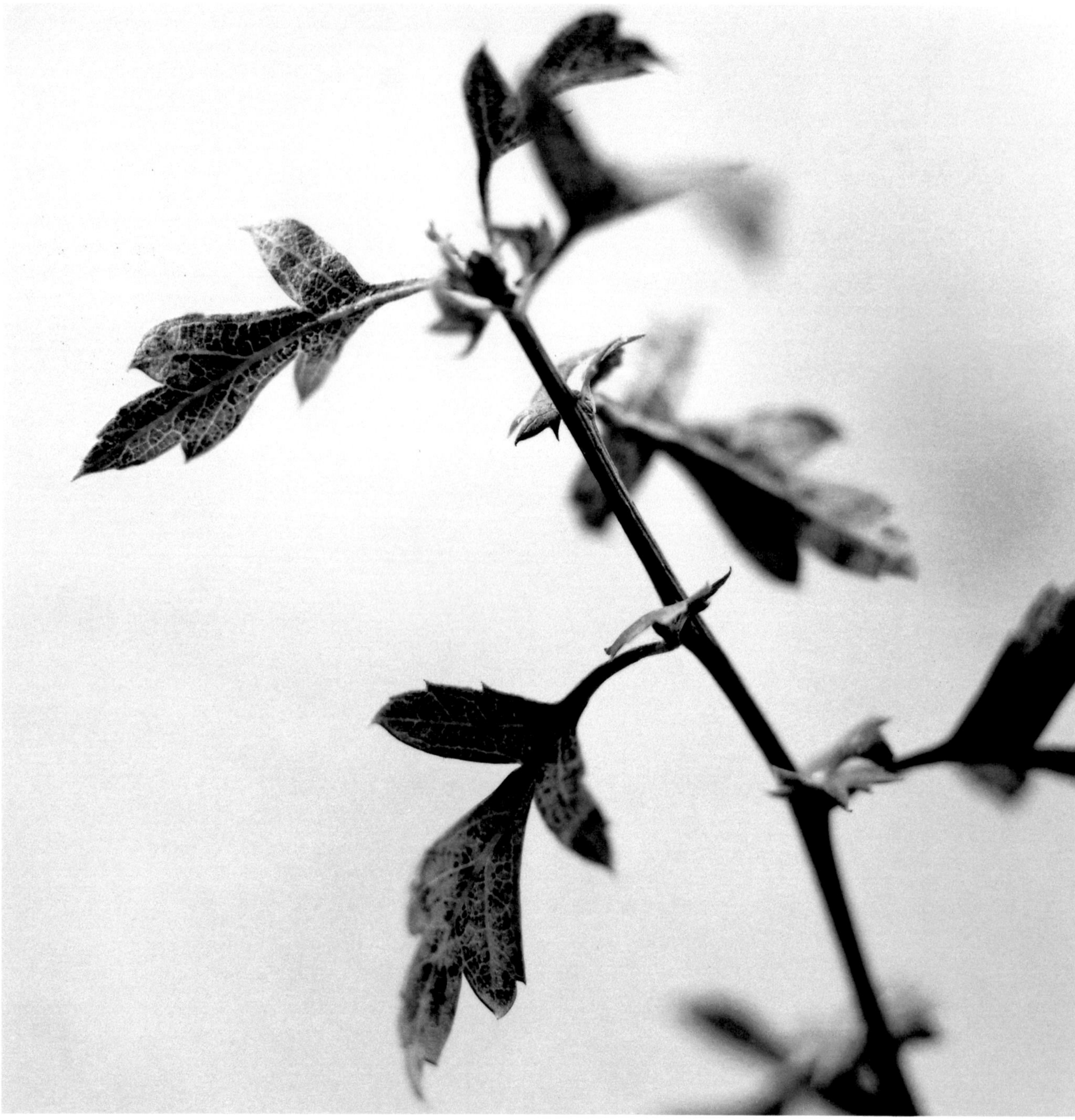

Crataegus monogyna

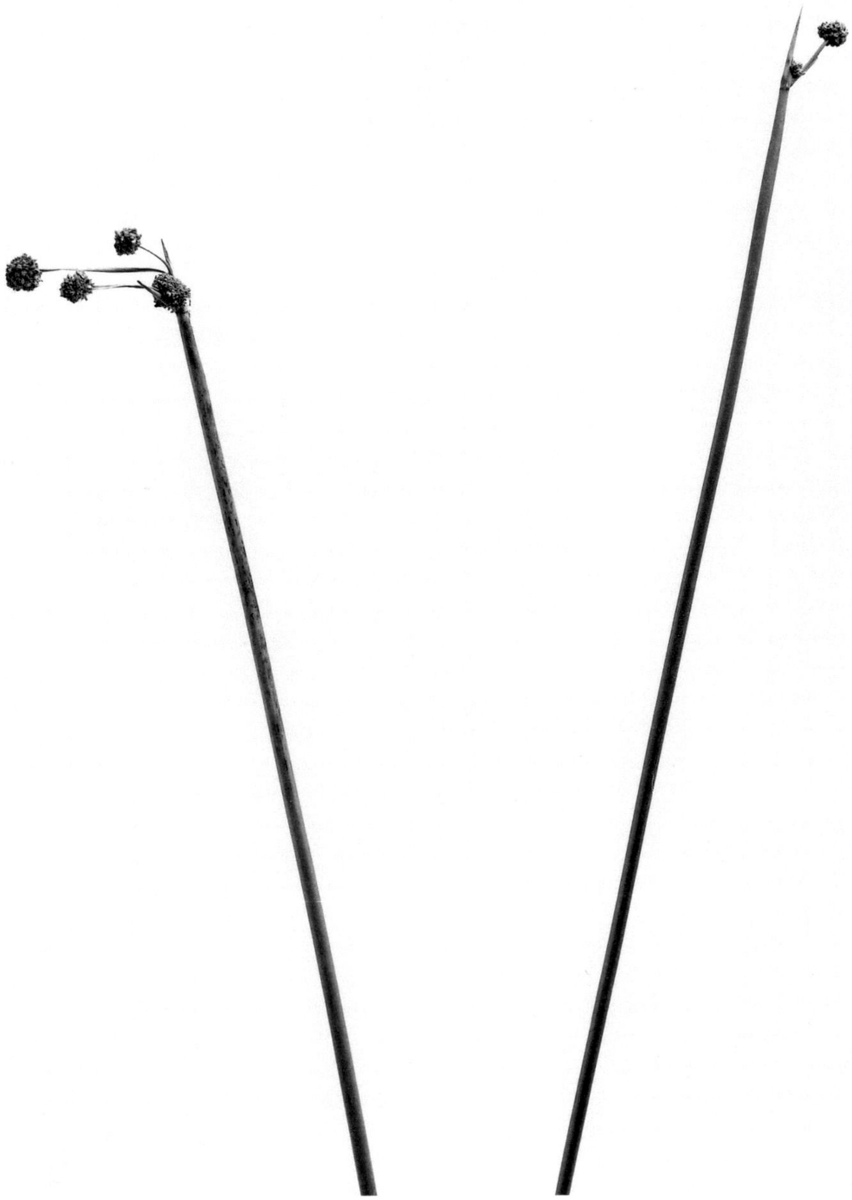

Scirpoides holoschoenus

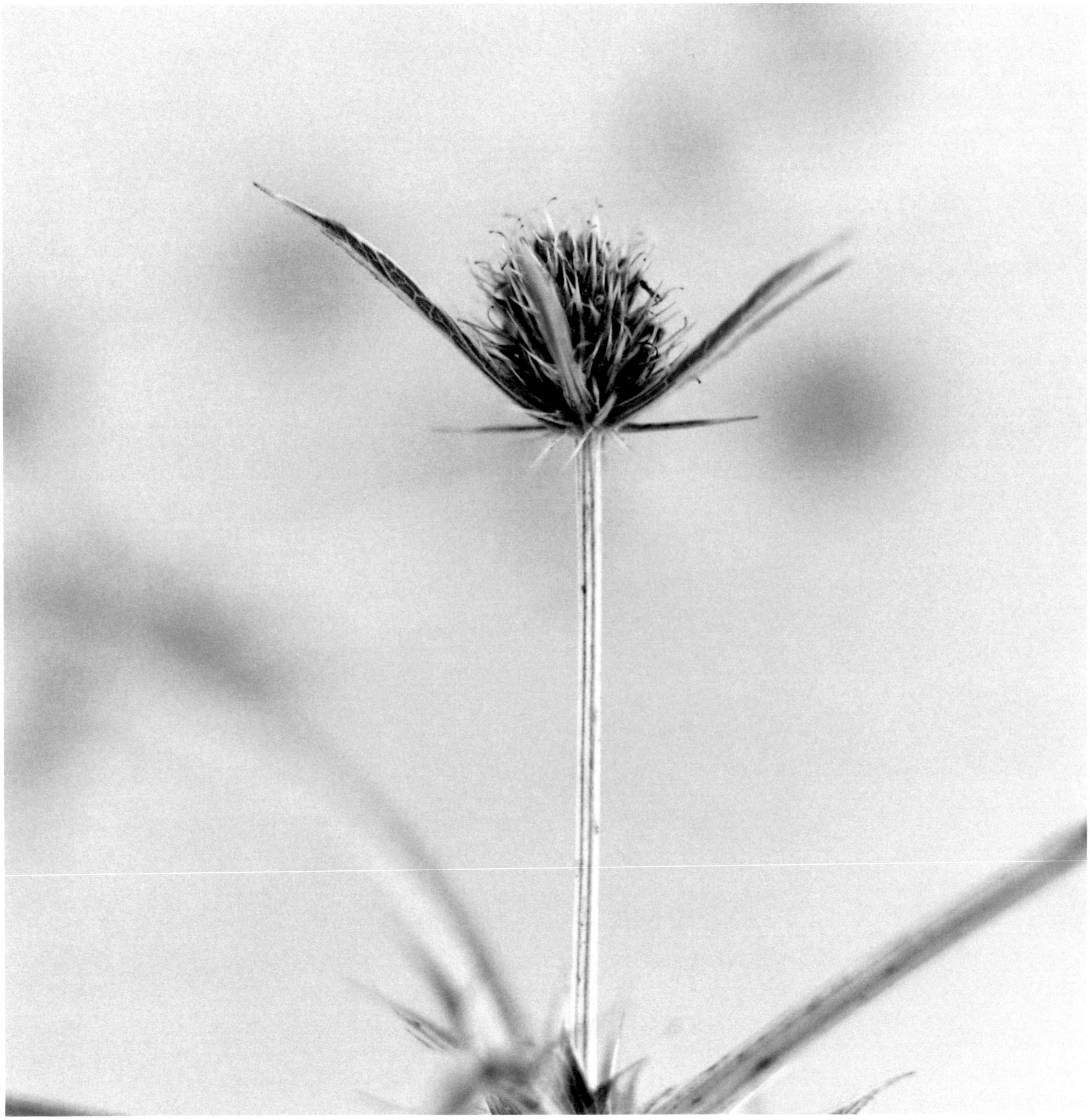

Eryngium campestre

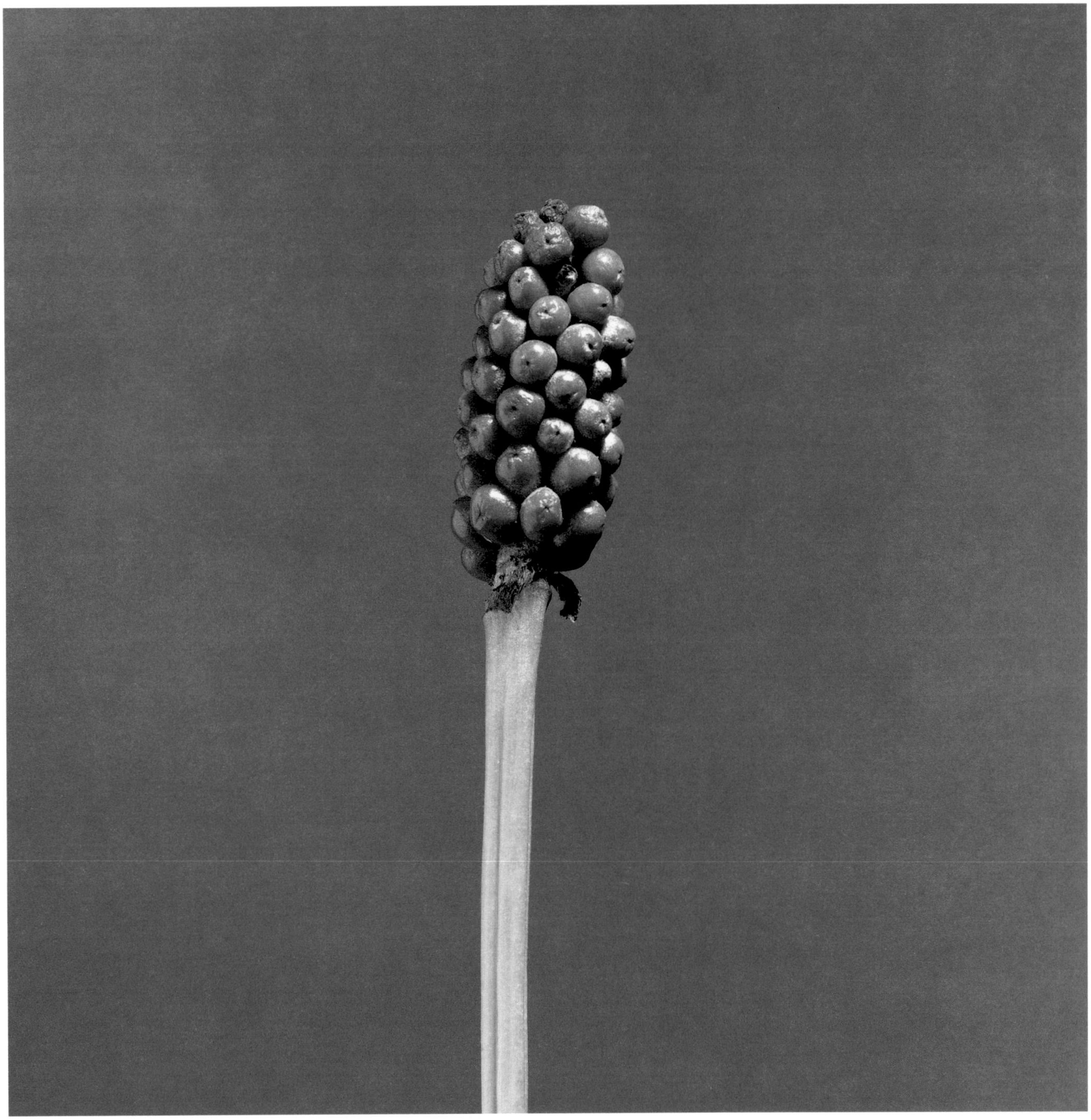

Arum italicum

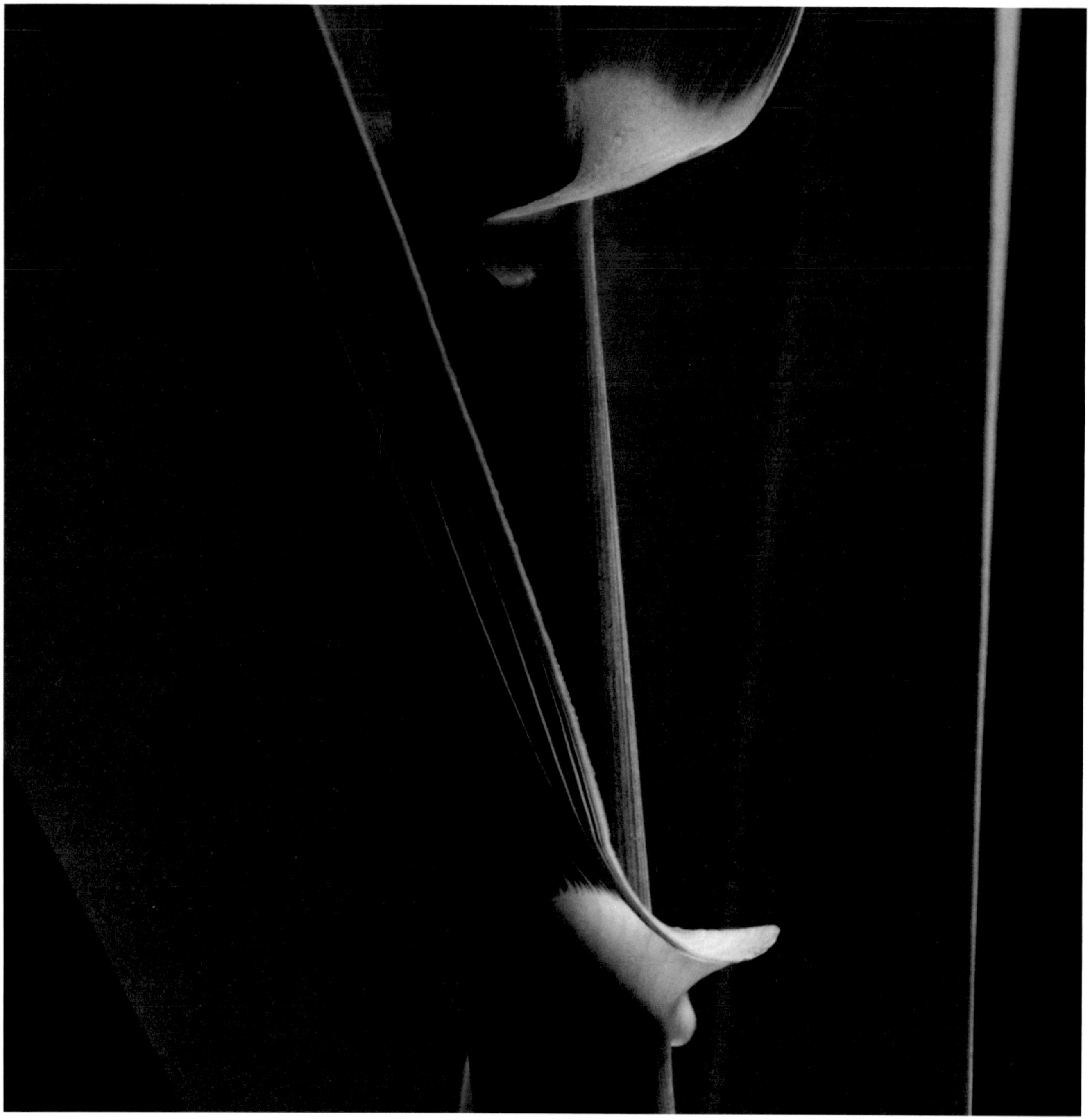

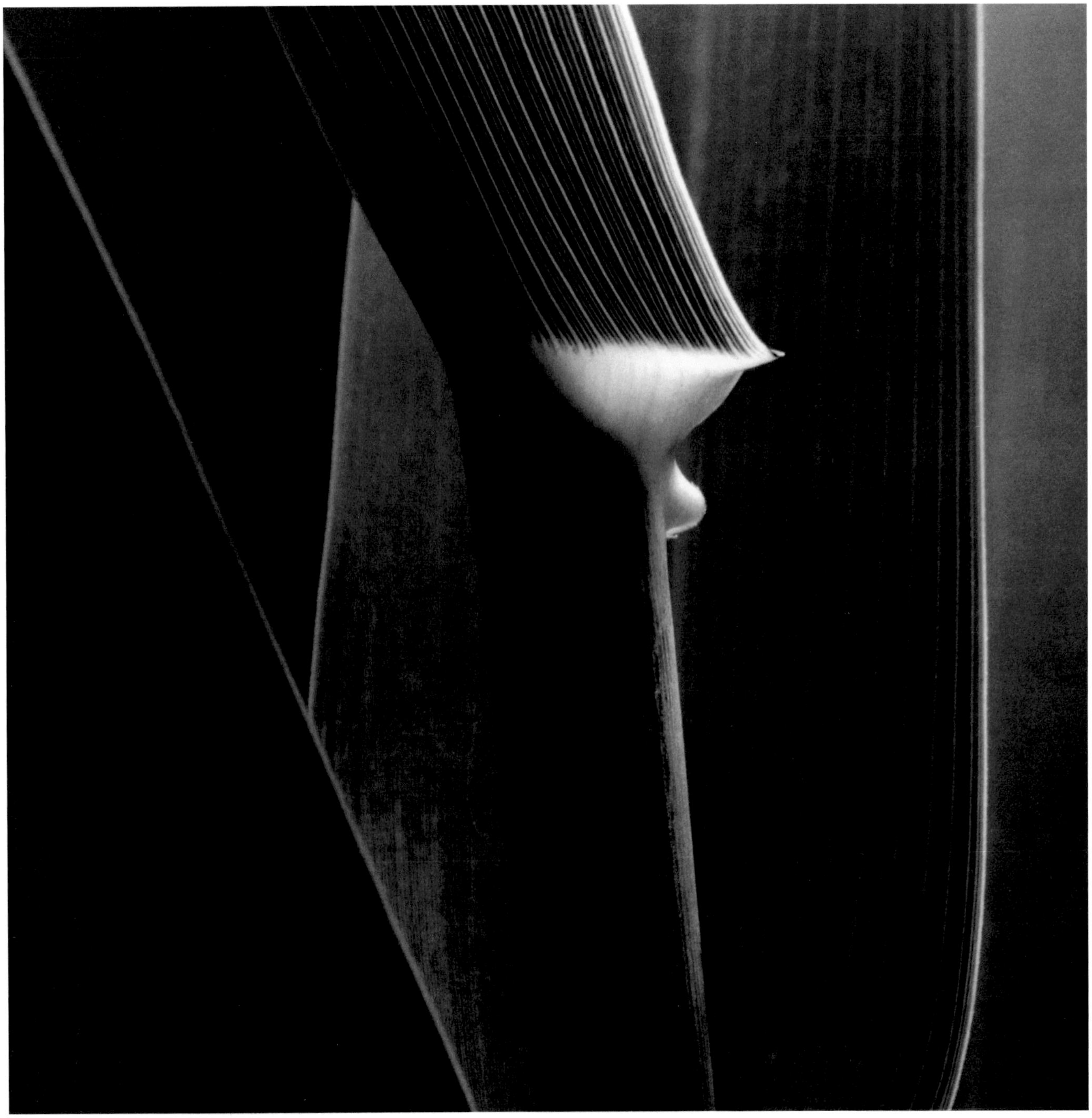

Arundo donax

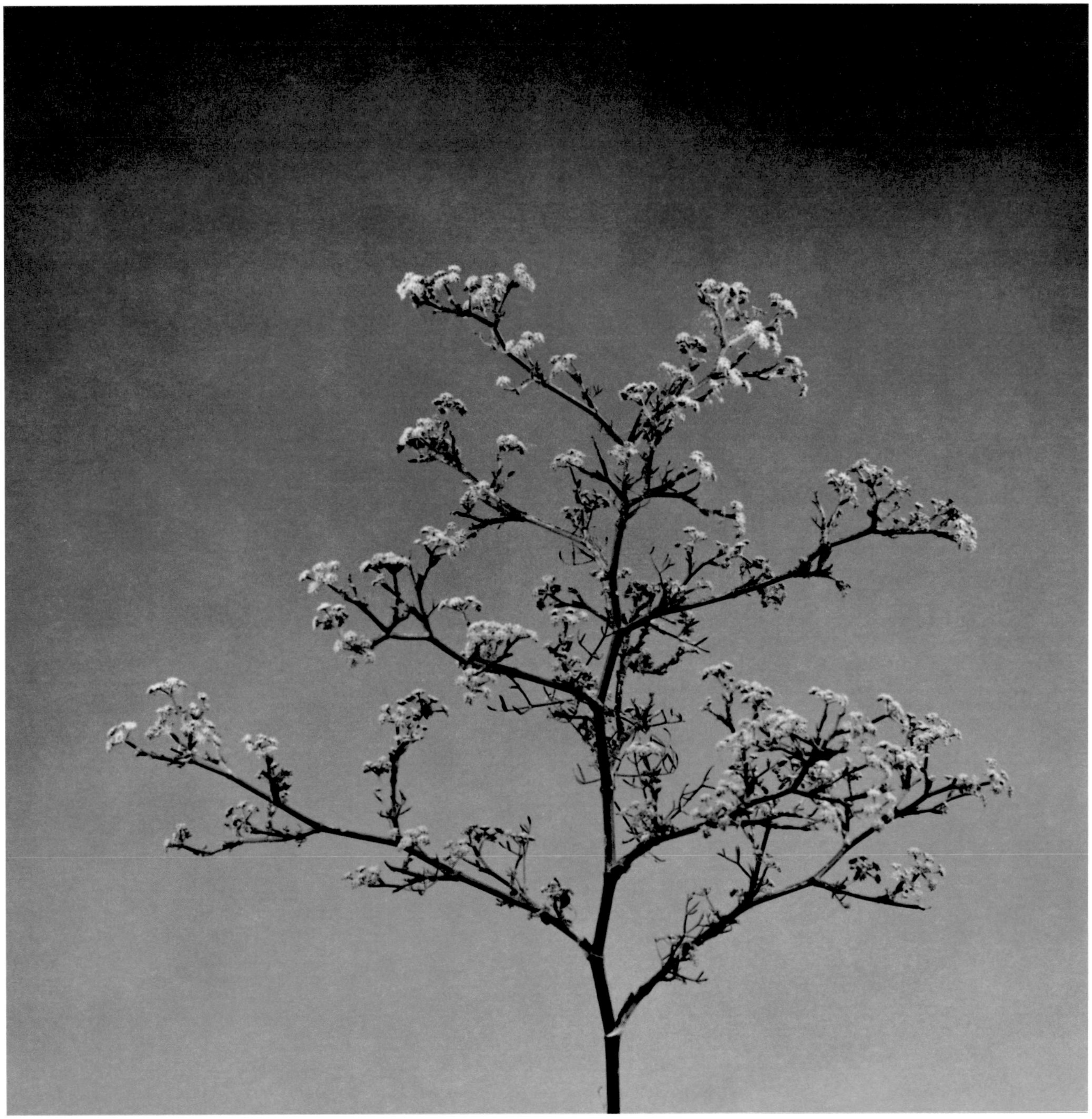

Seseli tortuosum

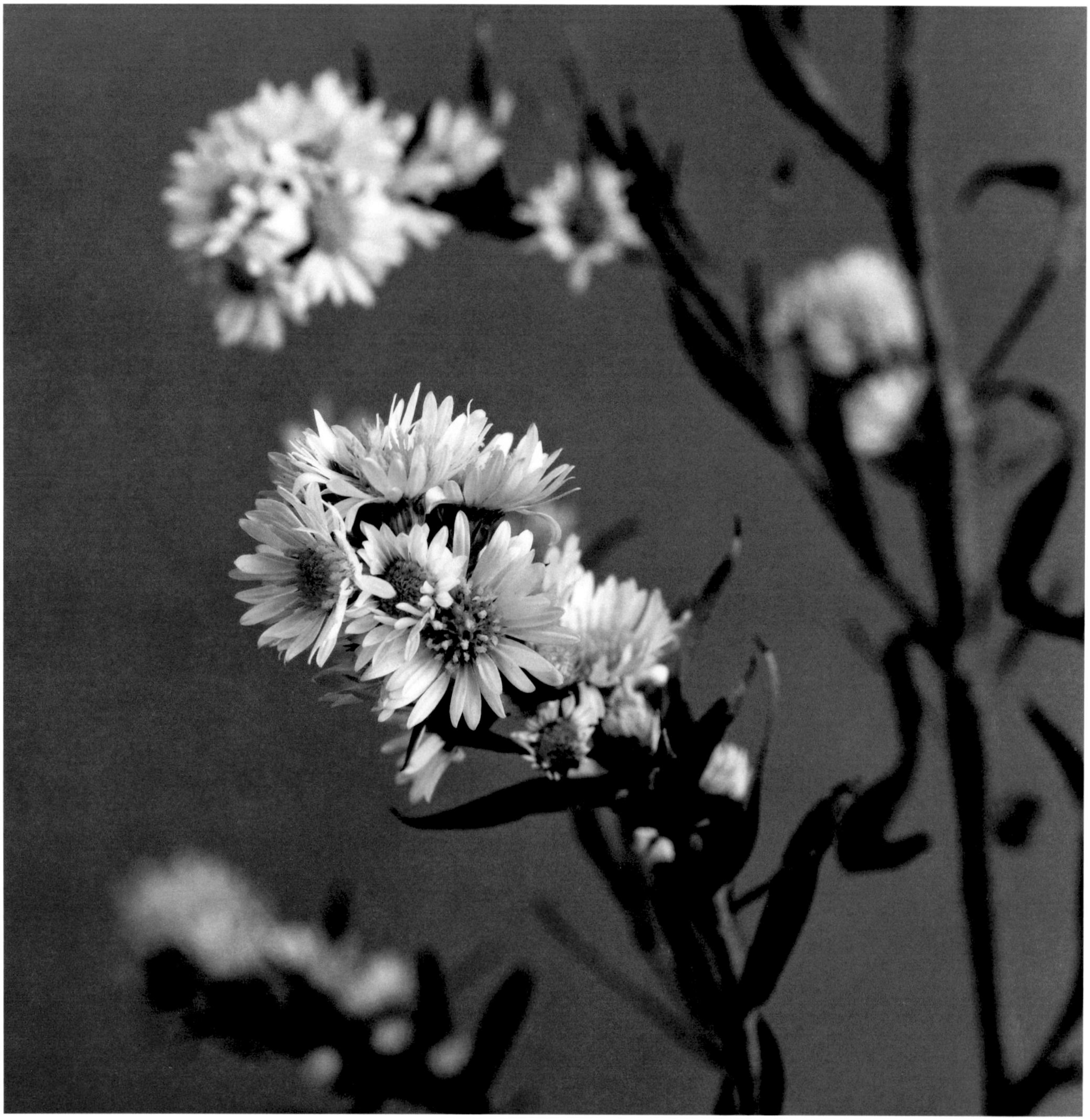

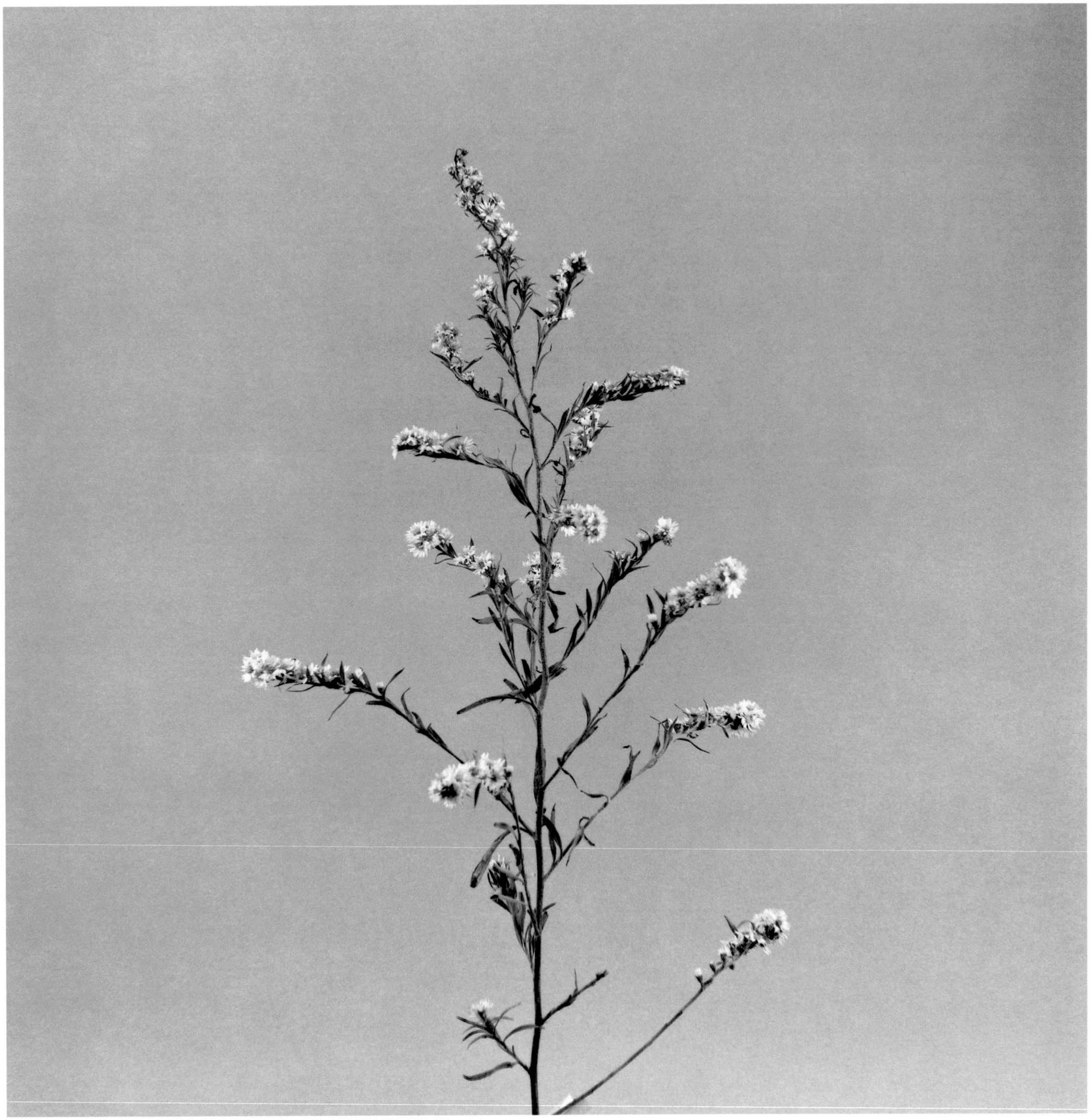

Aster sedifolius

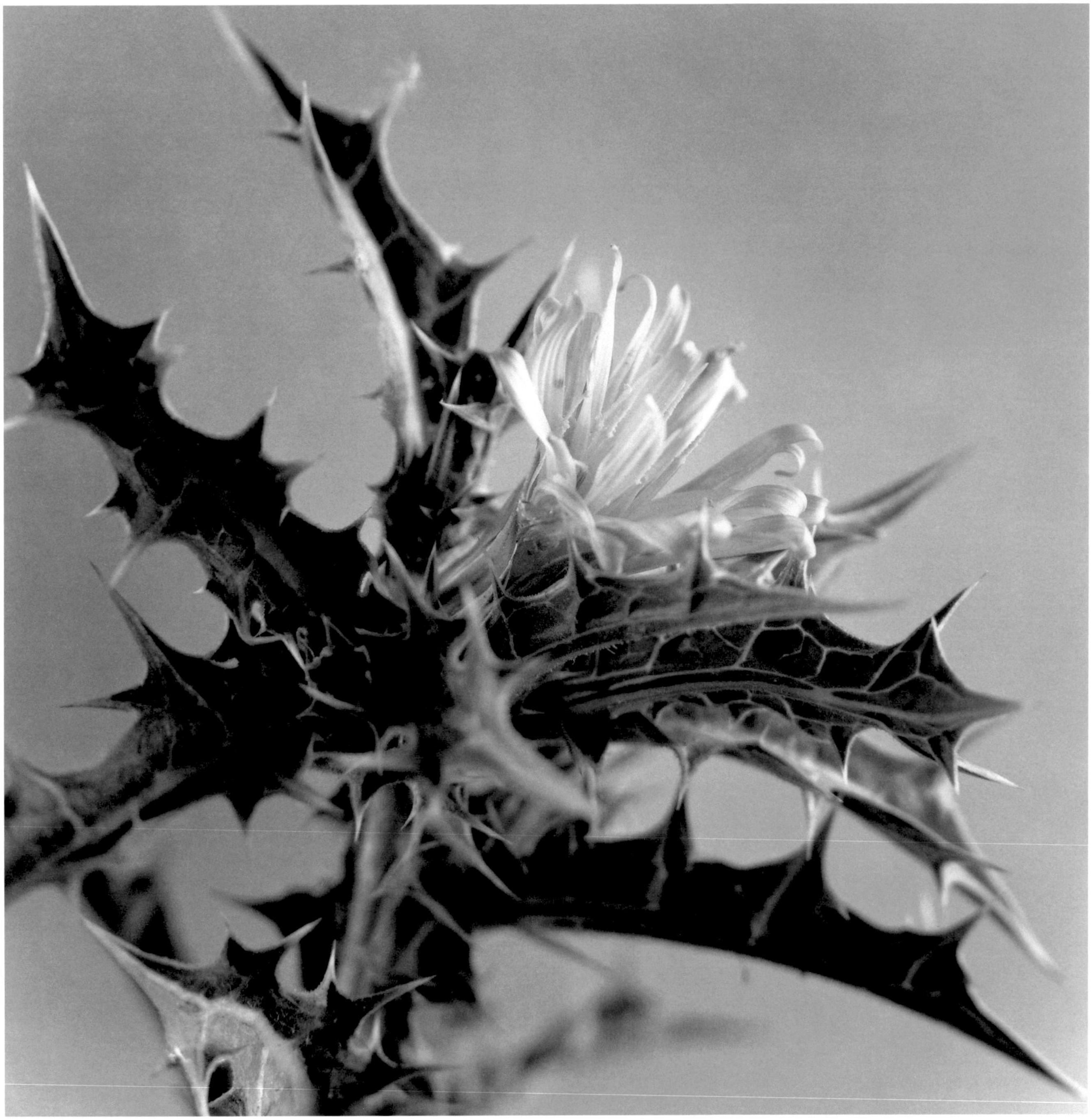

Scolymus hispanicus

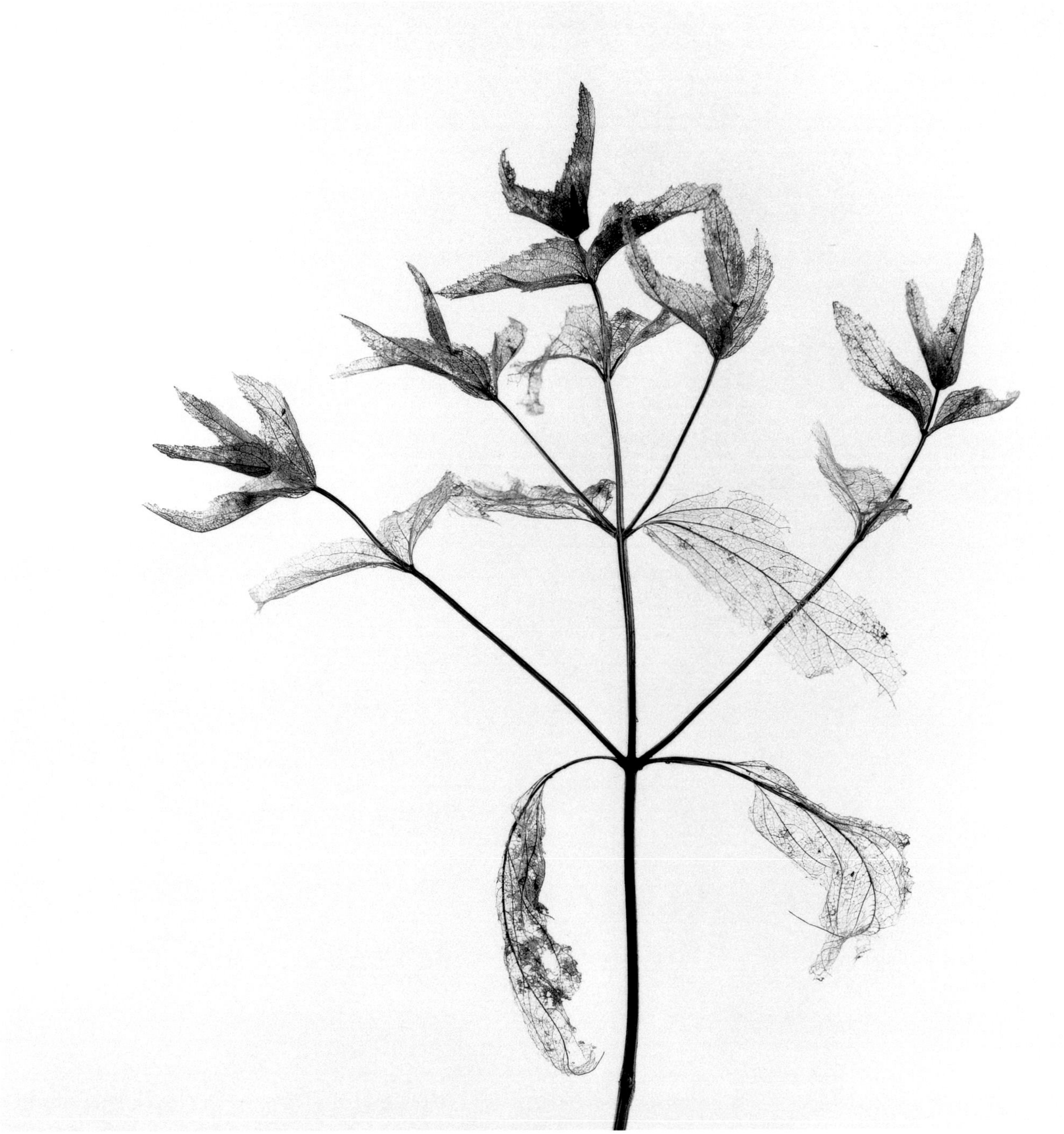

Phlomis herba-venti

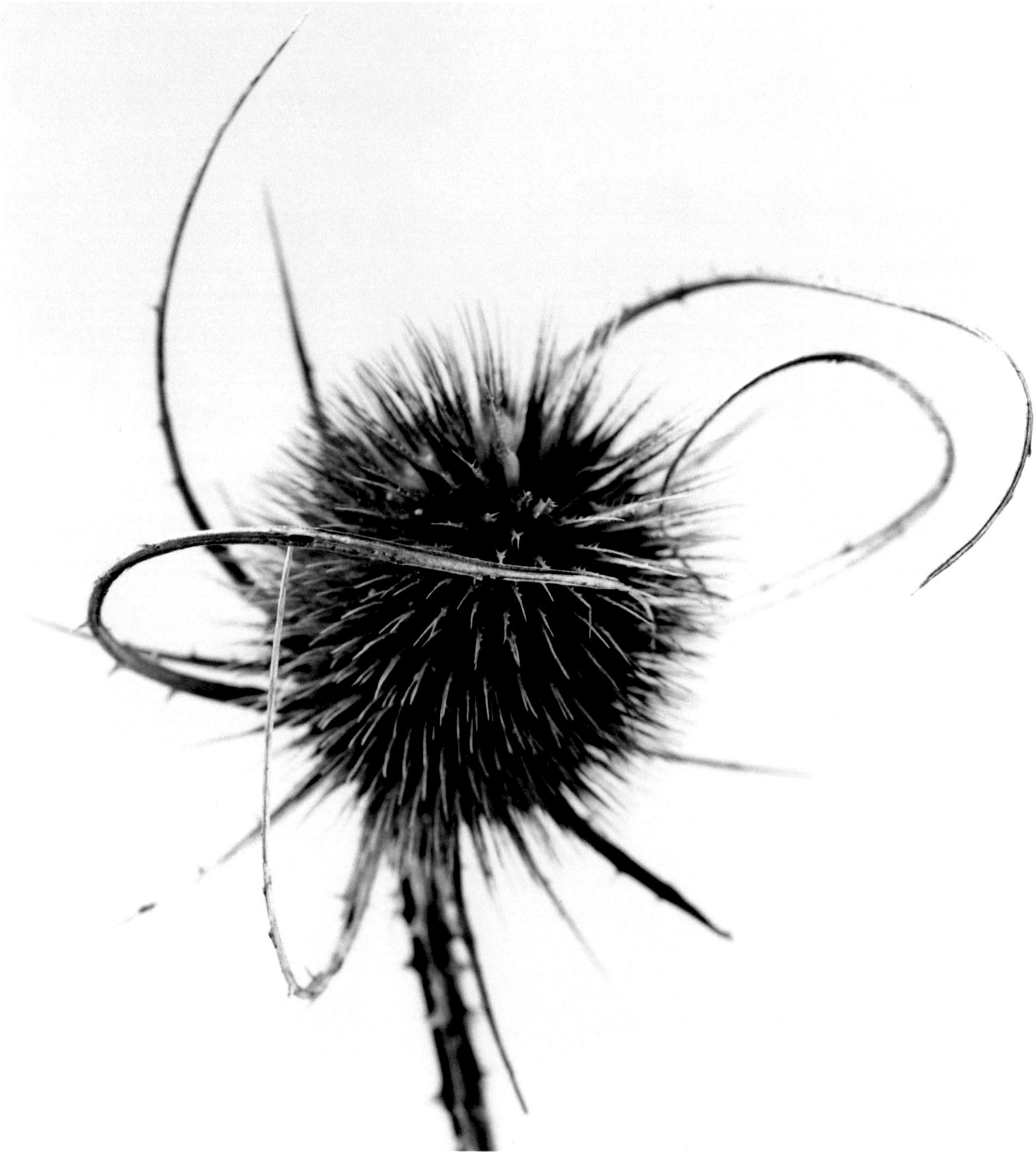

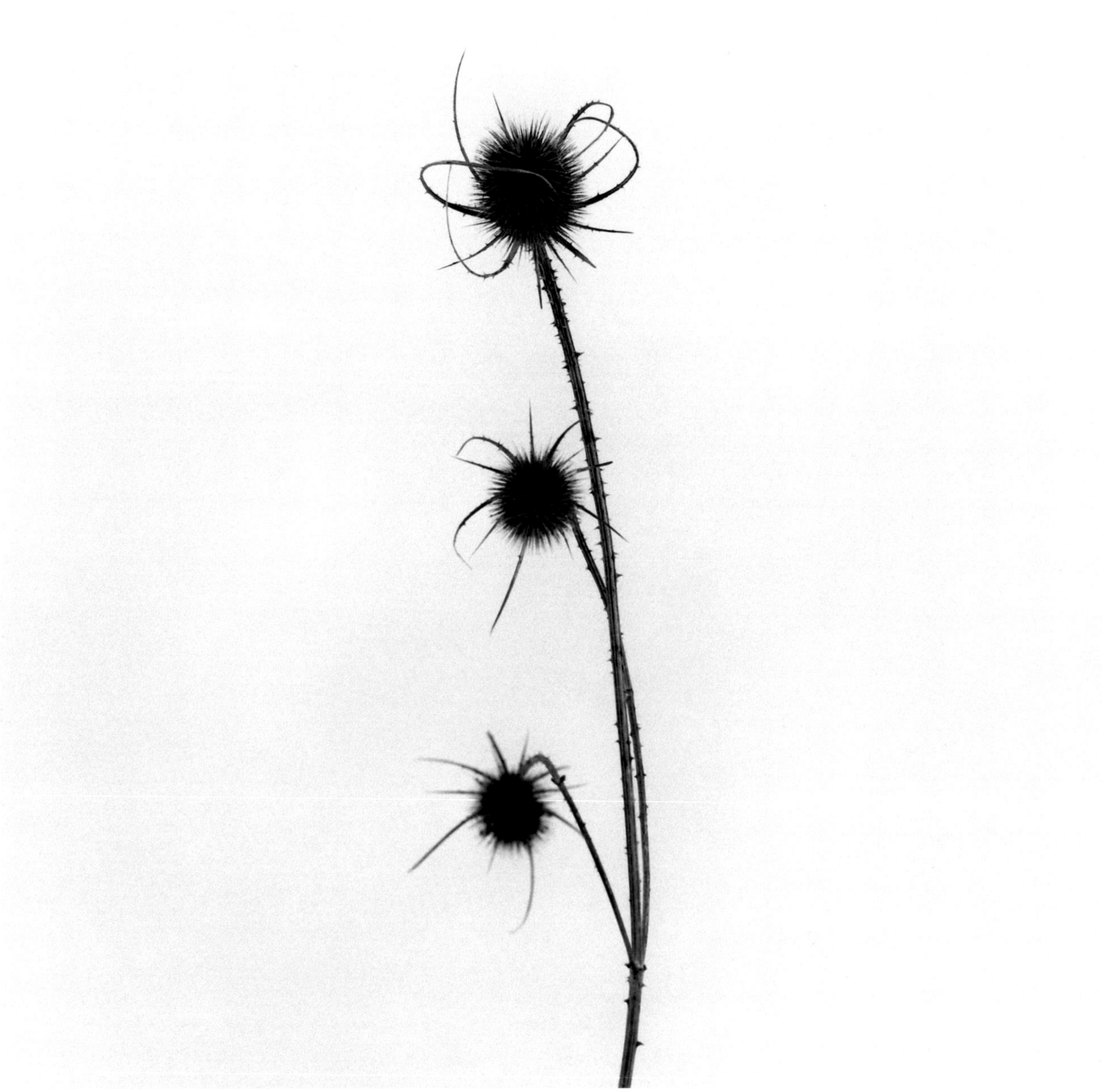

Dipsacus fullonum

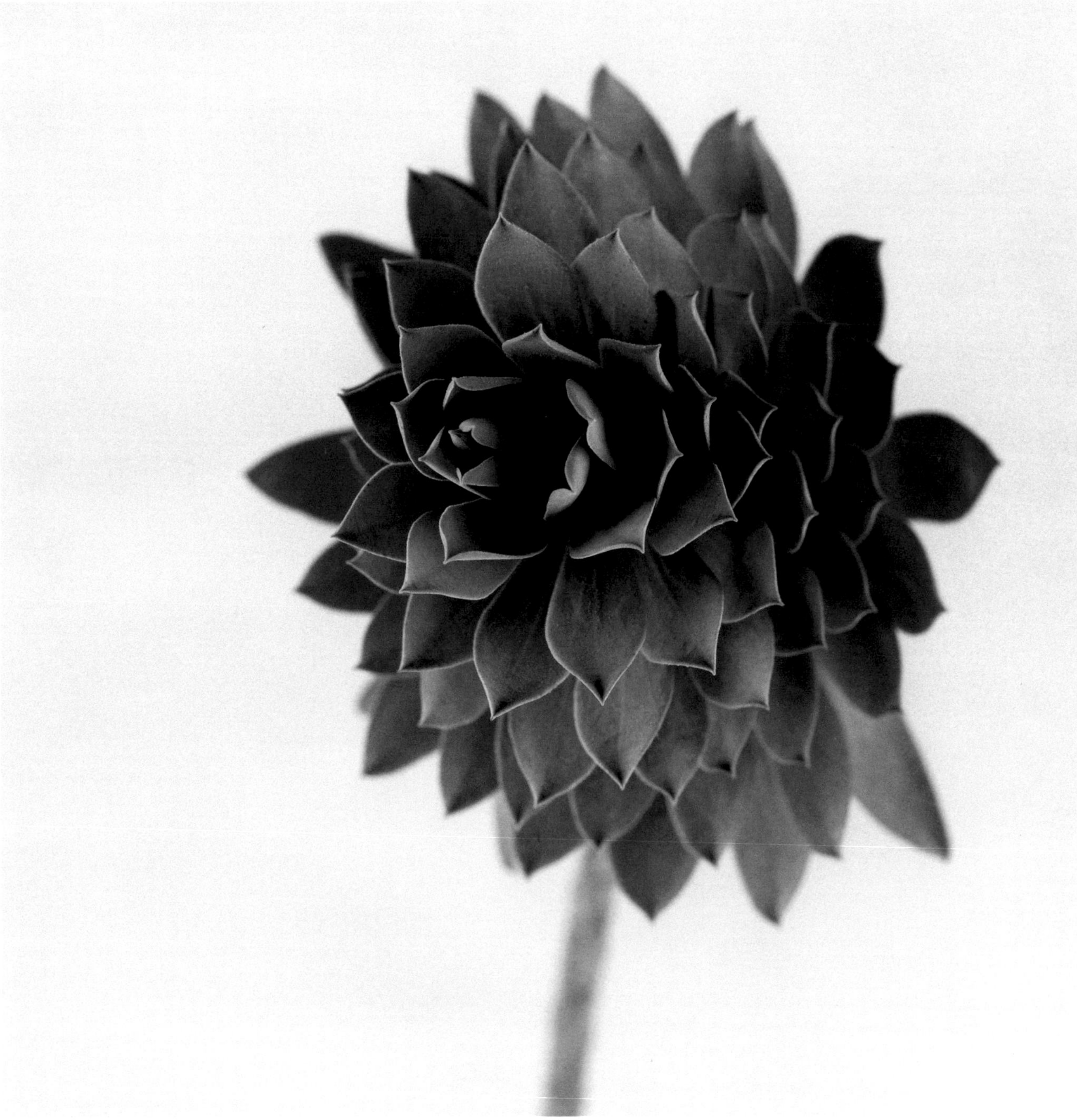

Euphorbia sp.

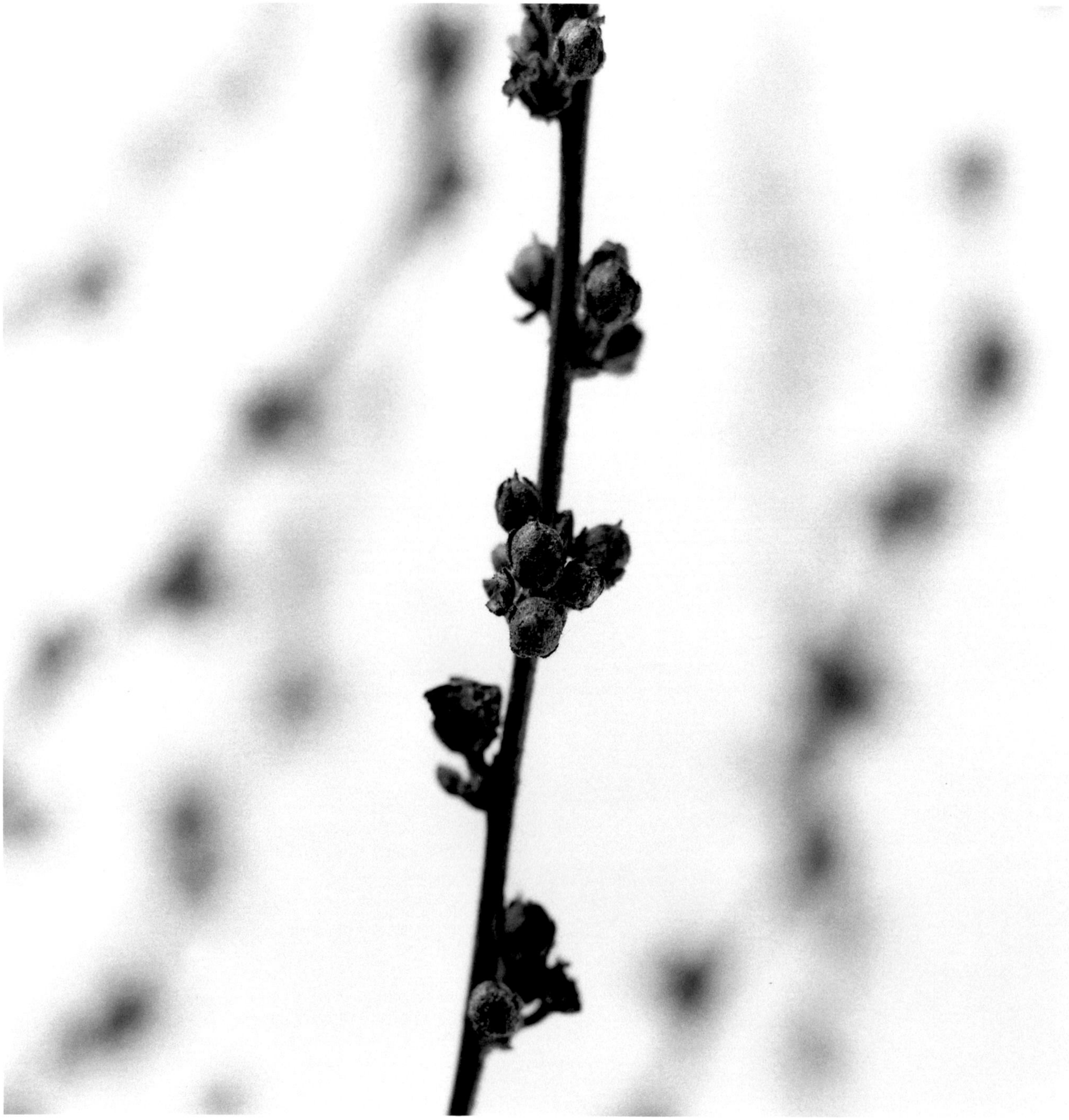

Verbascum sinuatum

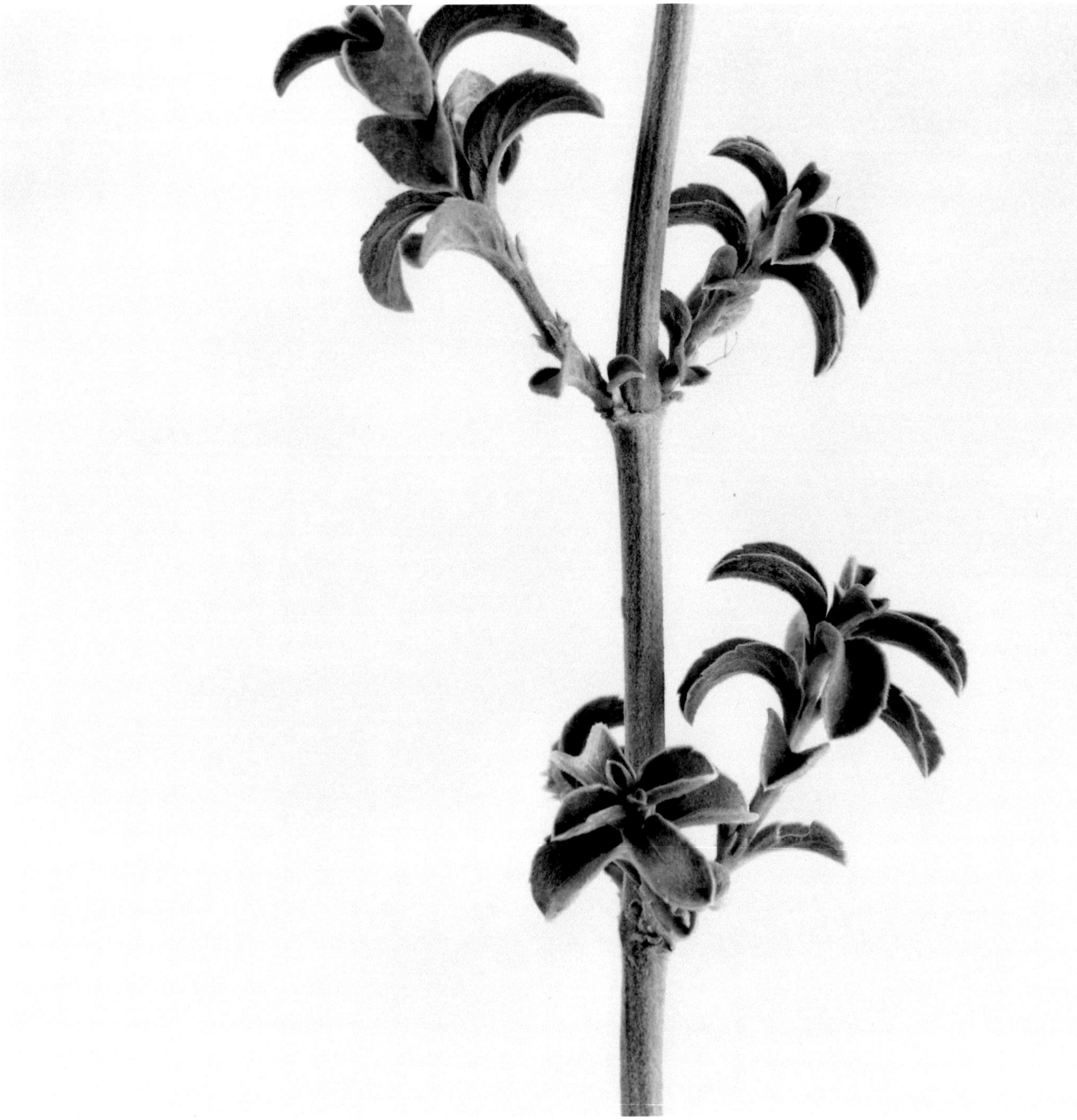

Sp. ign.

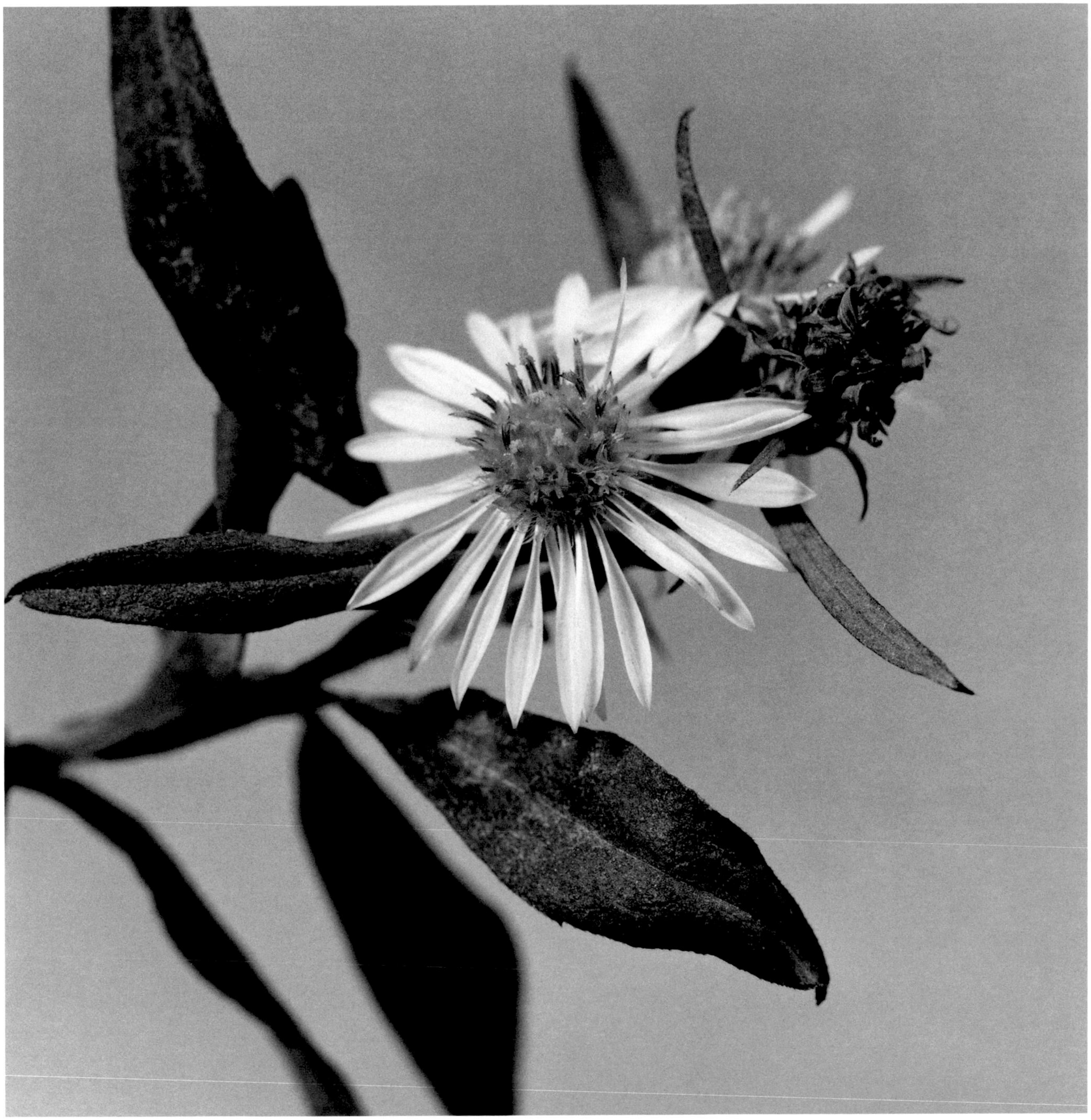

Bidens sp.

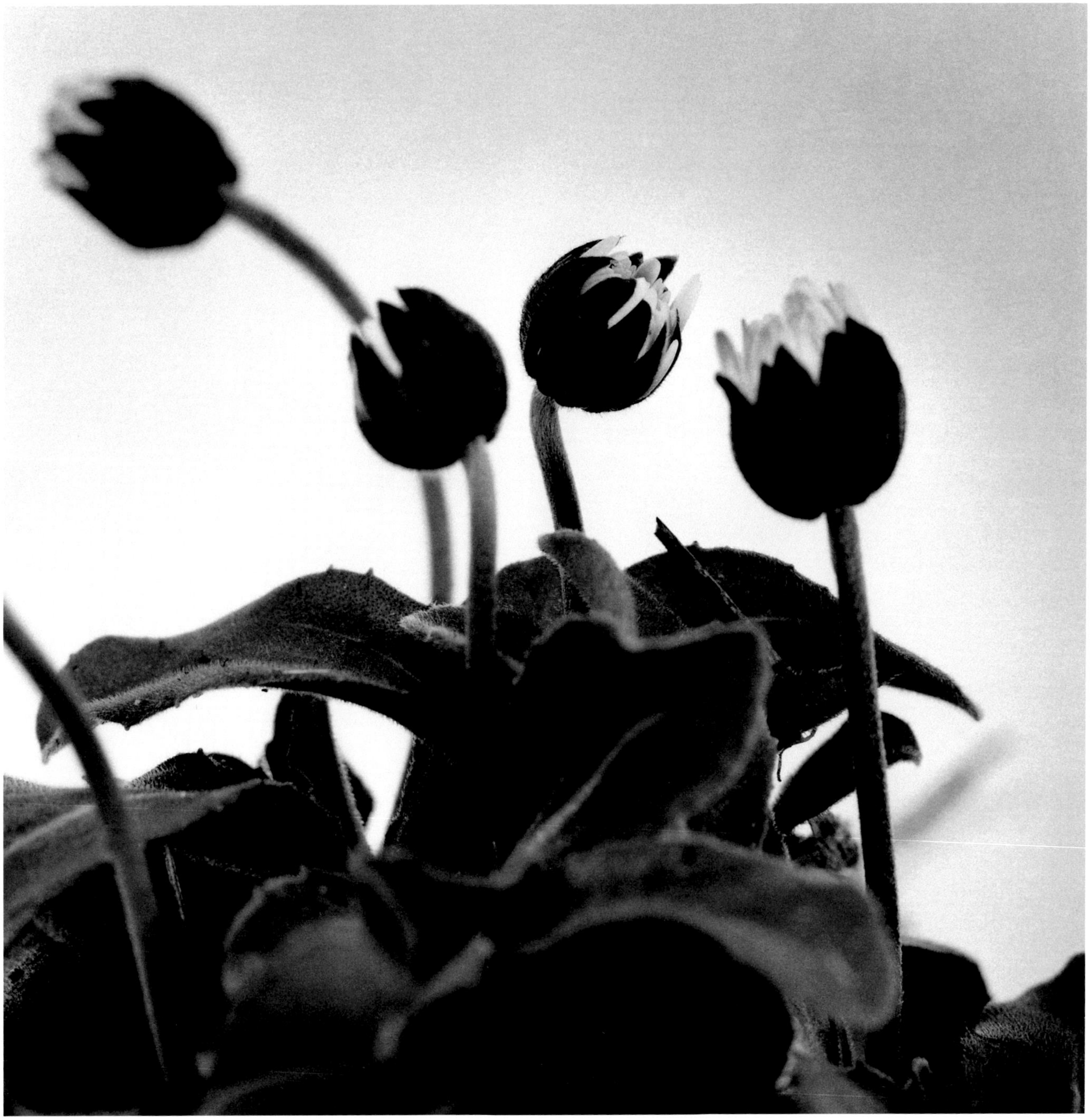

Bellis perennis

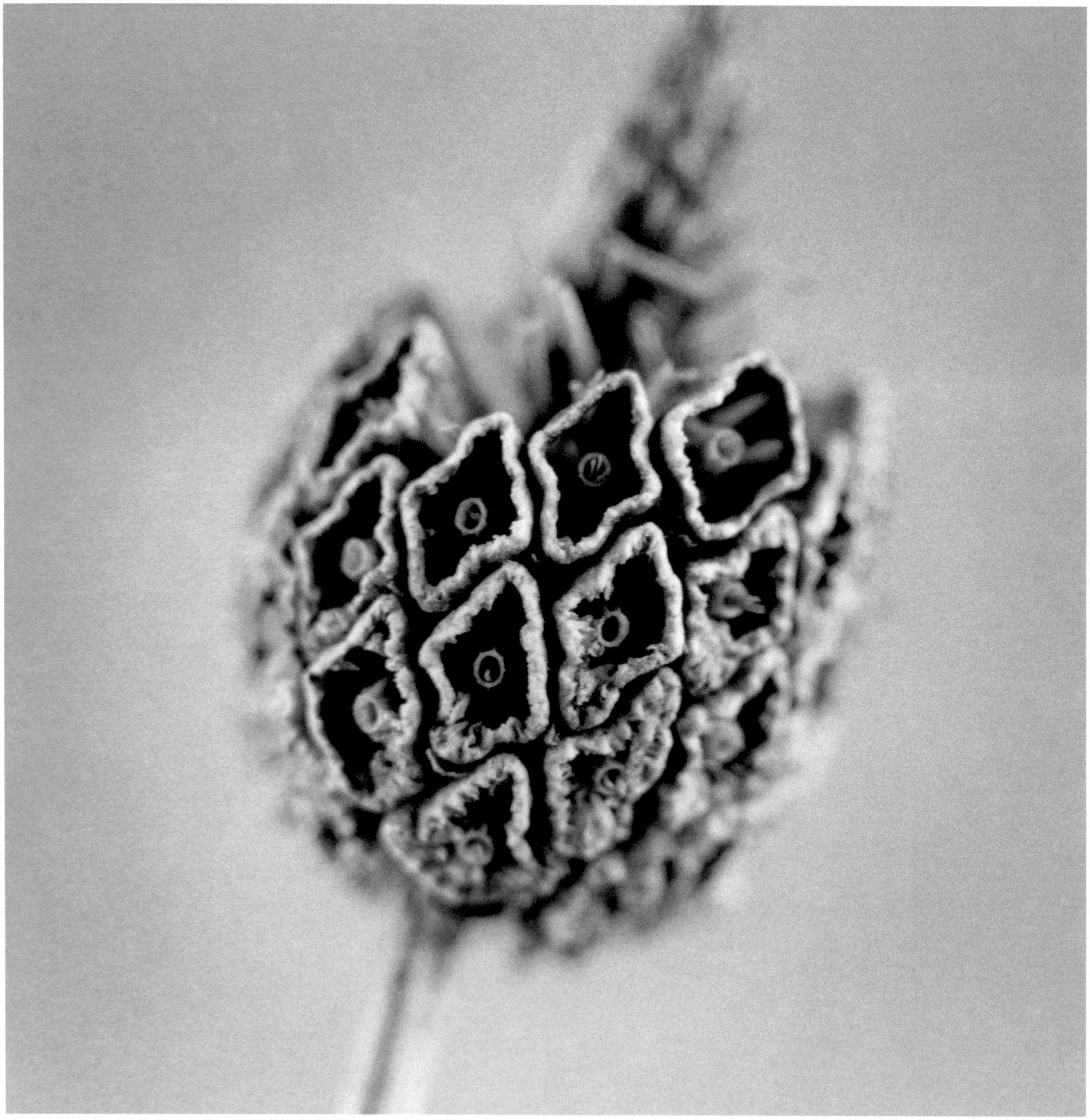

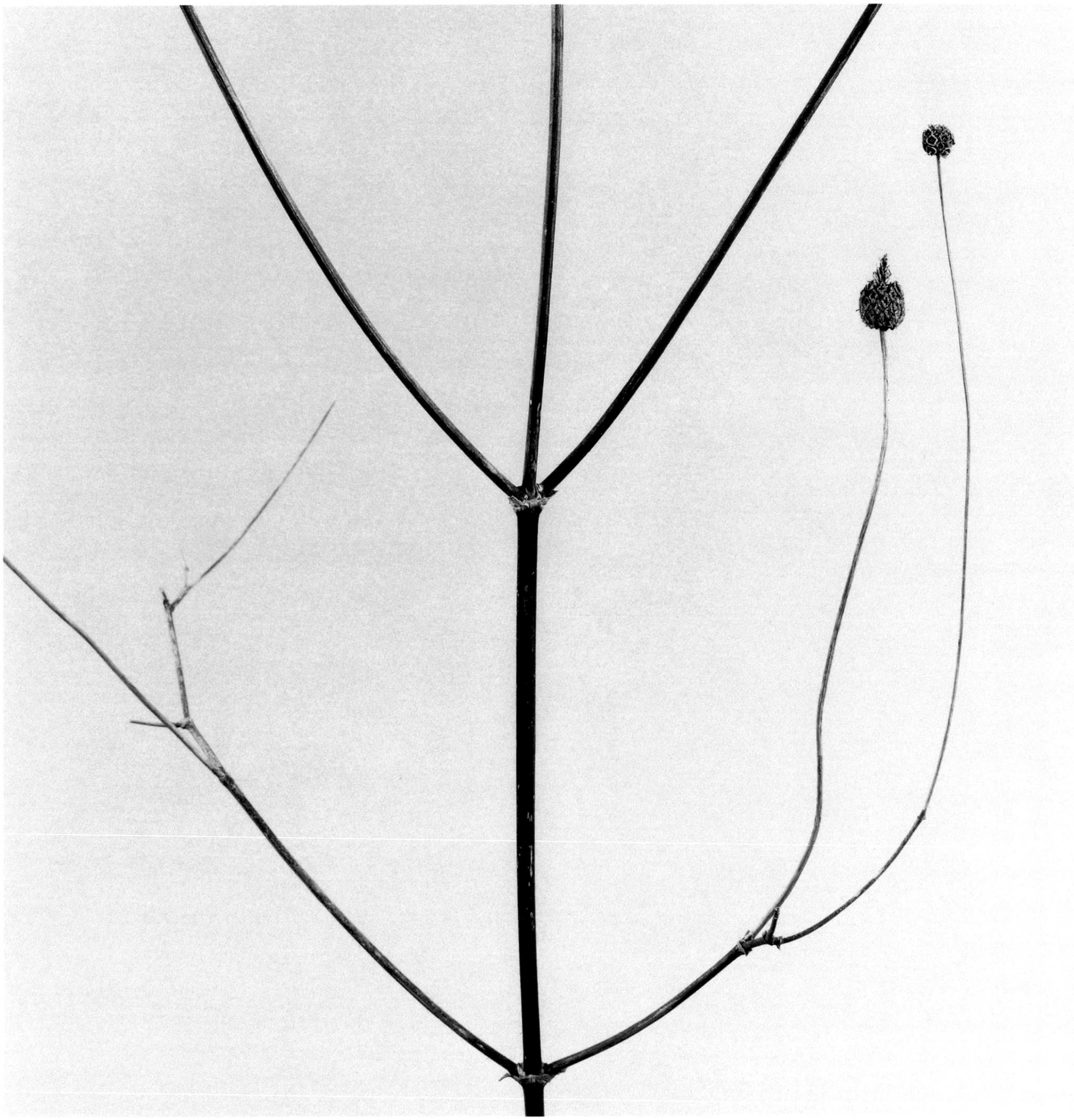

Scabiosa atropurpurea

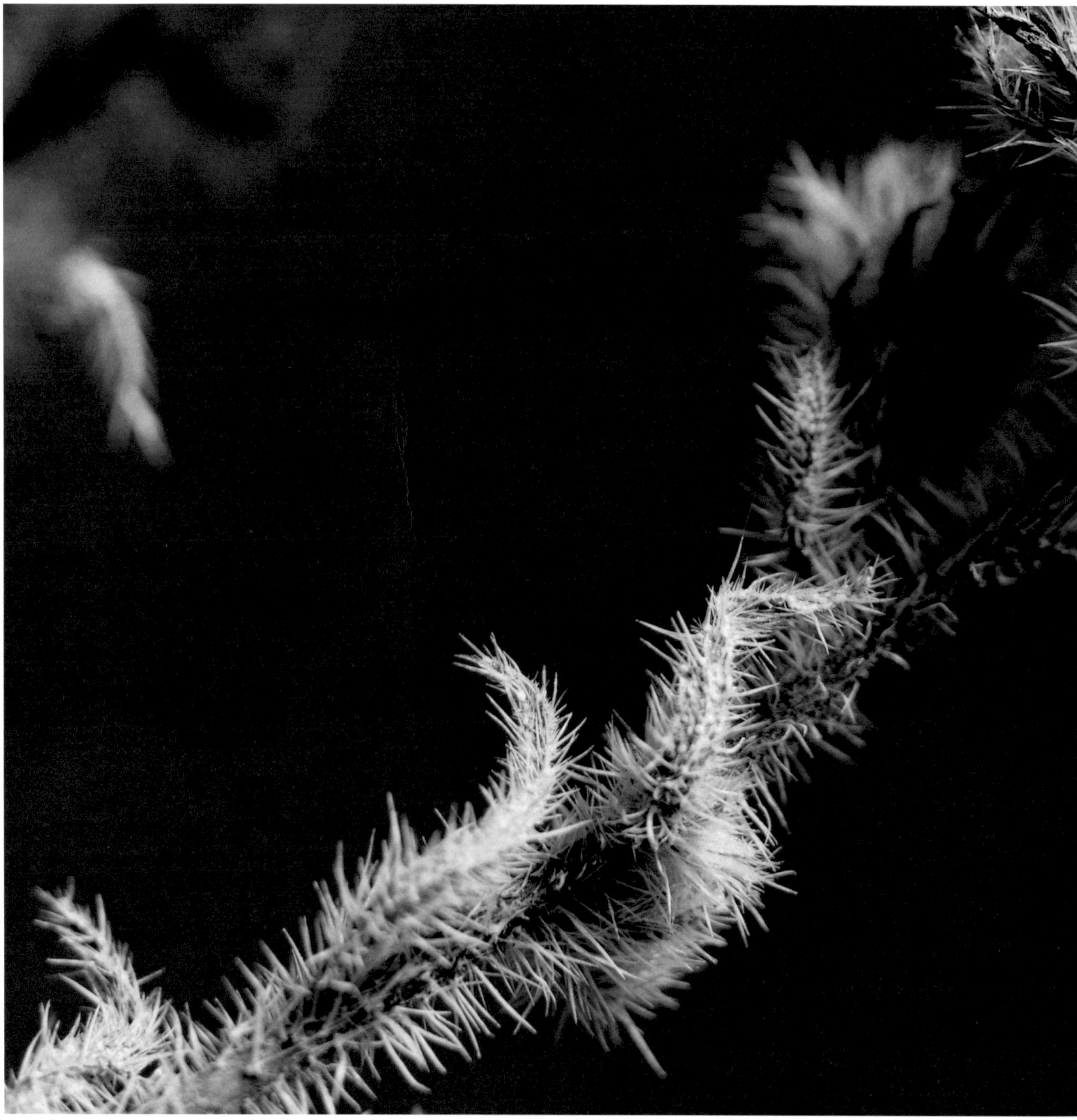

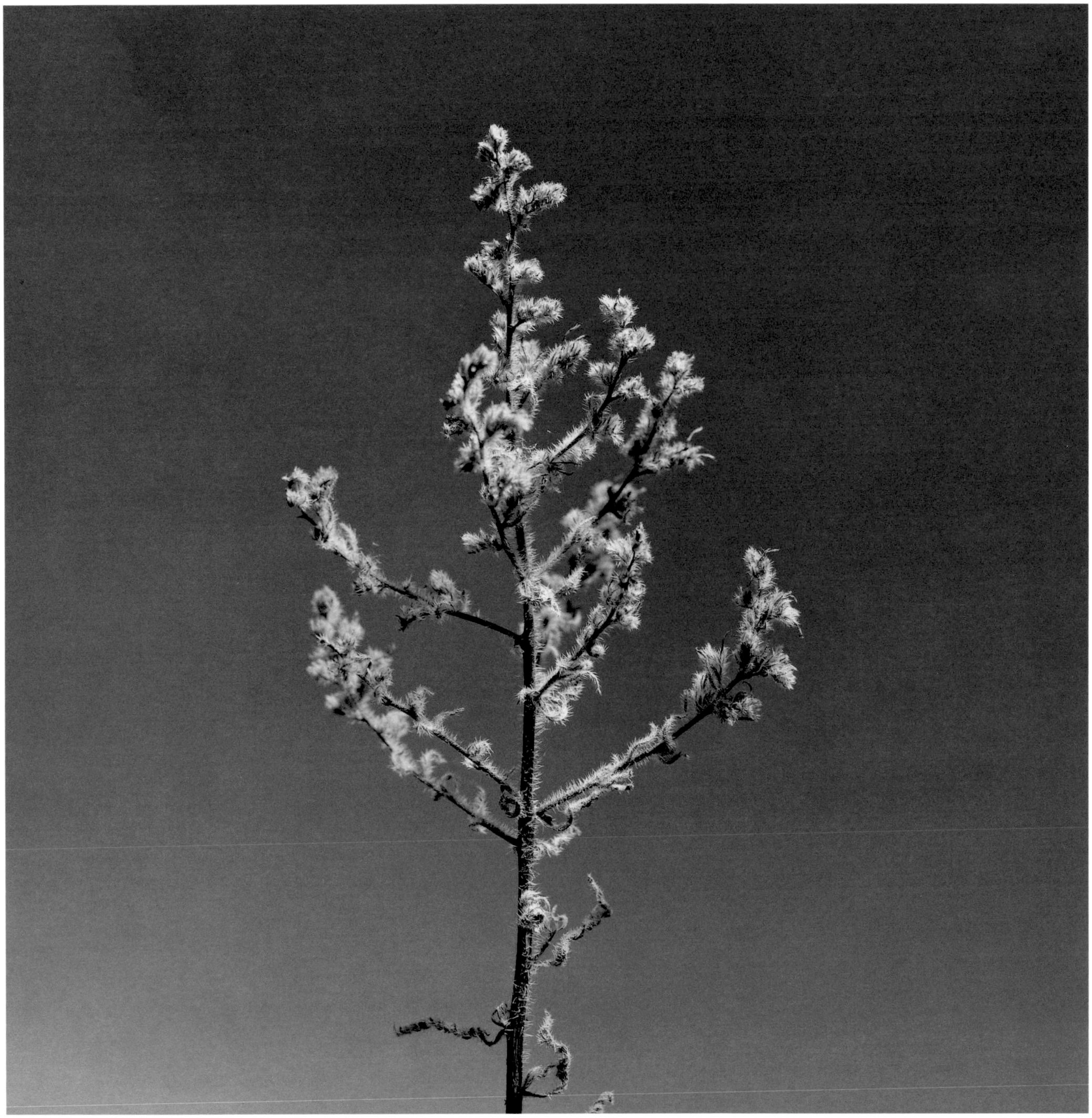

Echium italicum

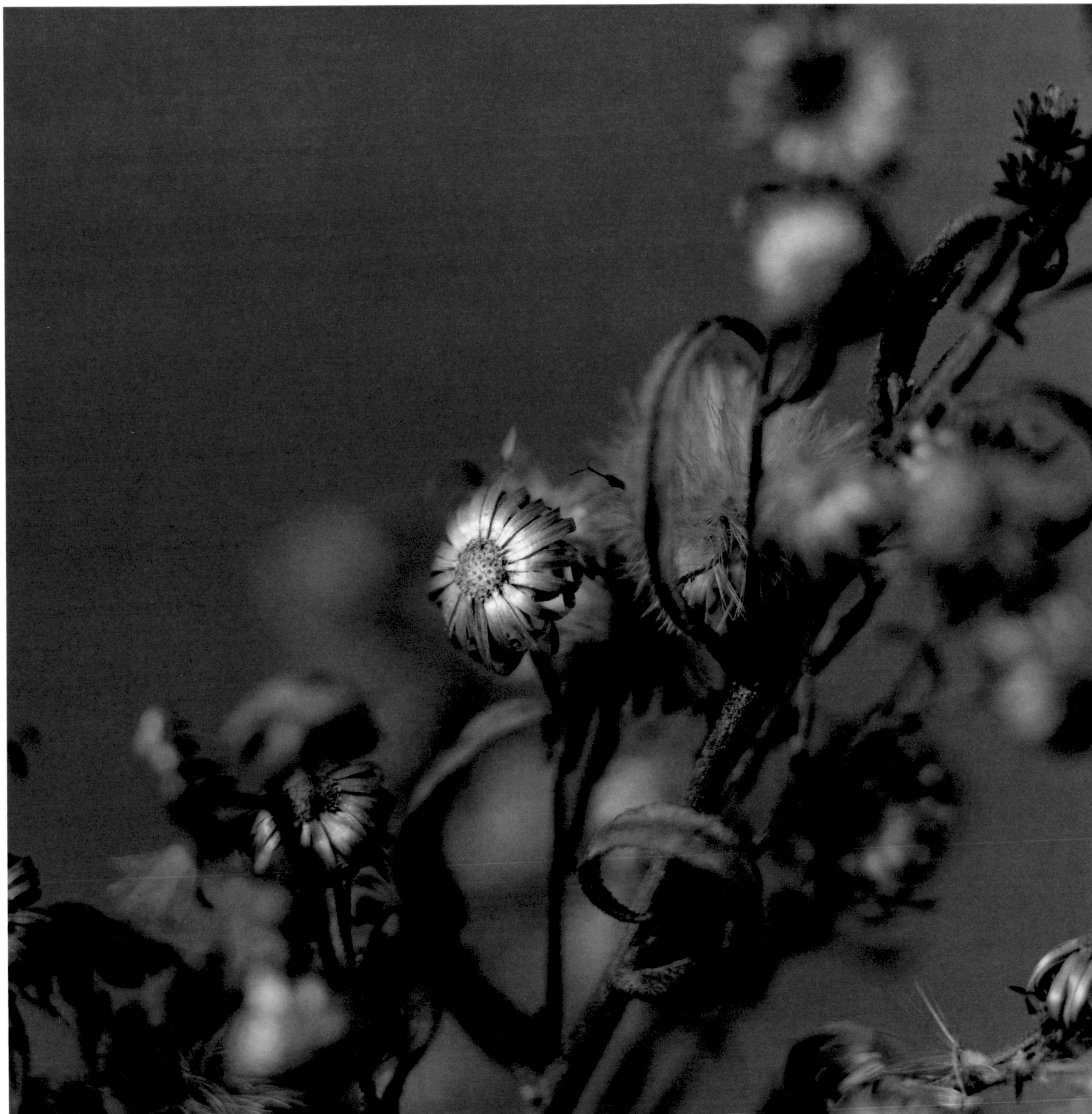

Dittrichia graveolens

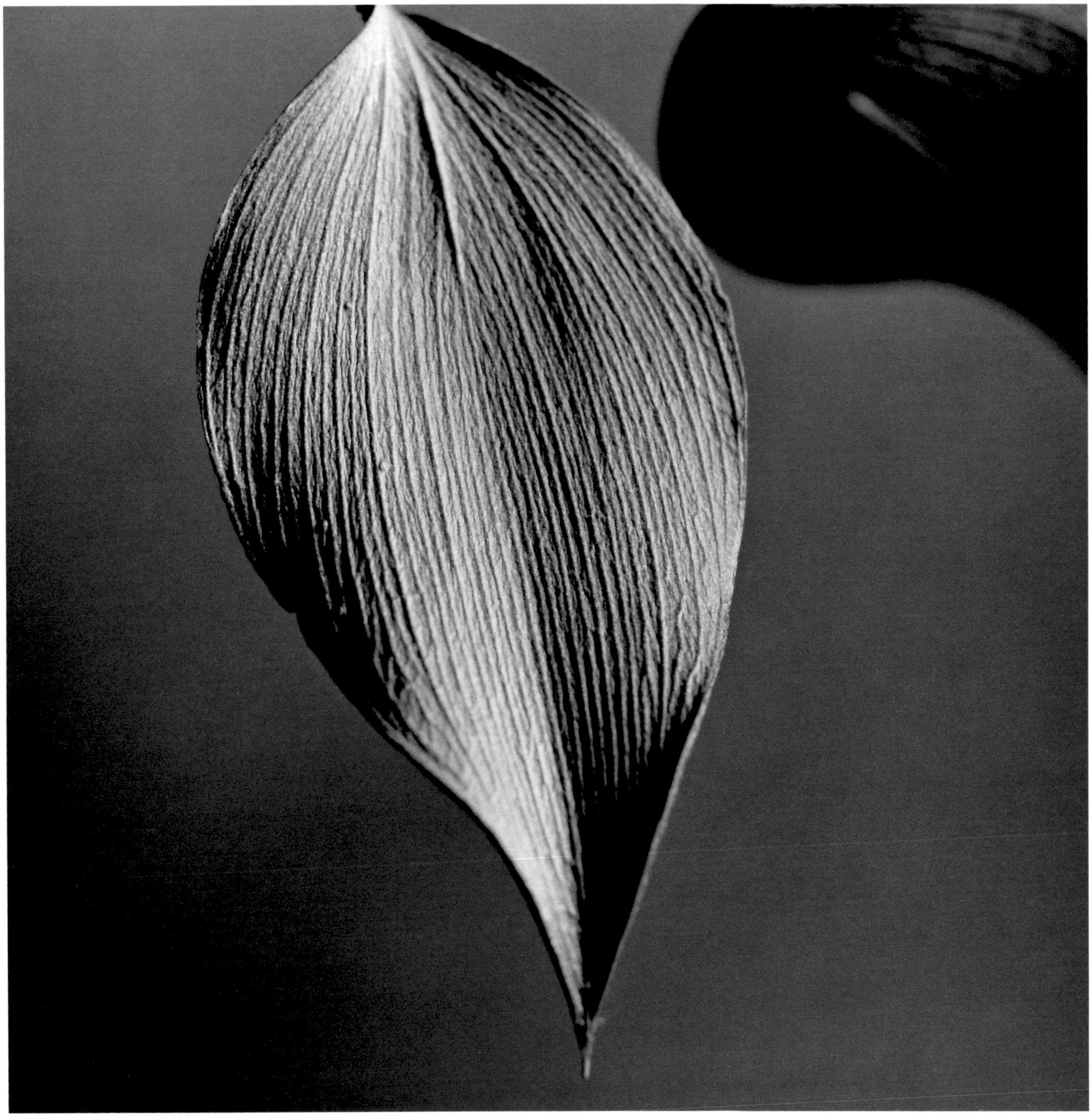

Ruscus hypoglossum

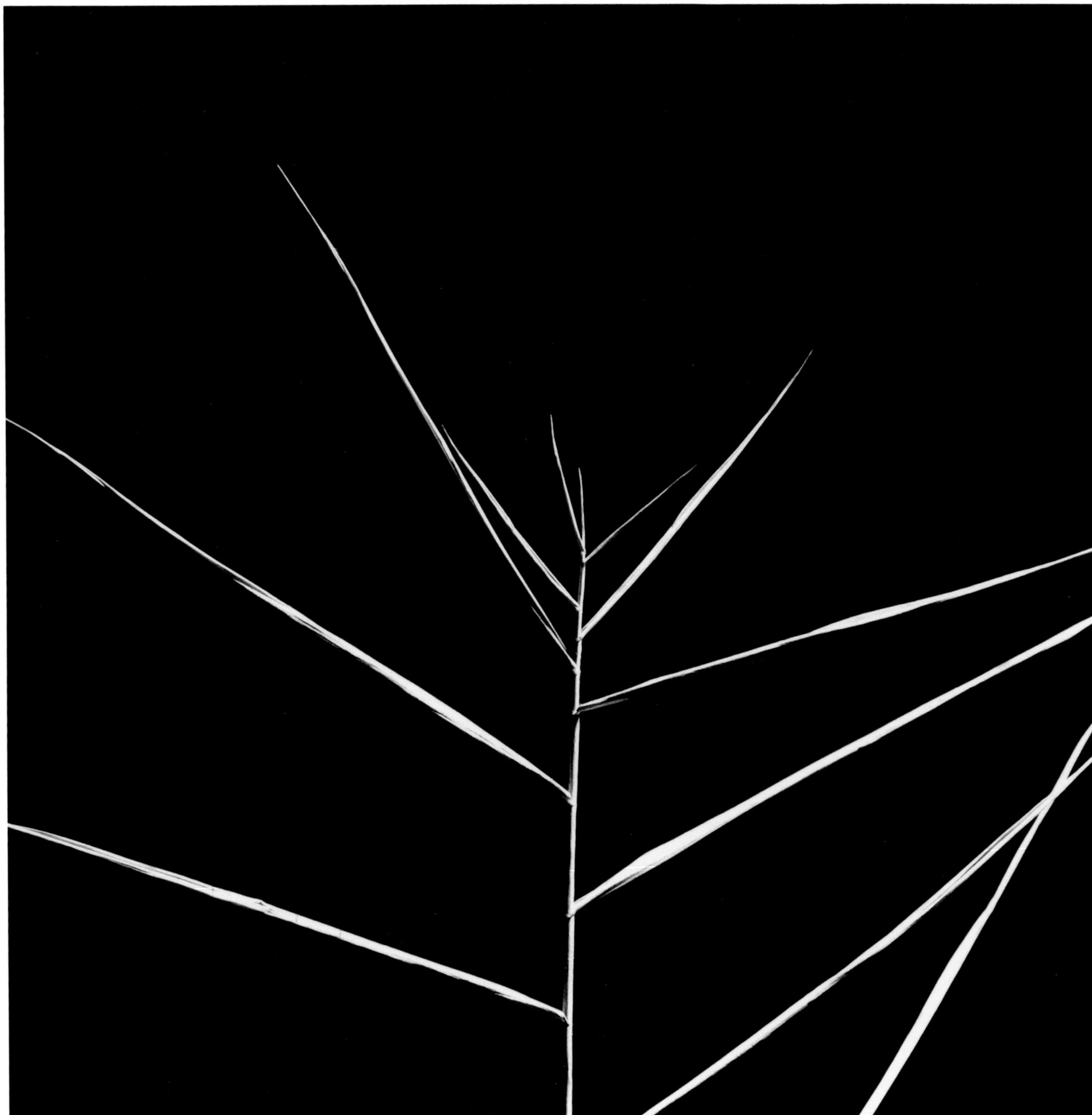

Phragmites australis

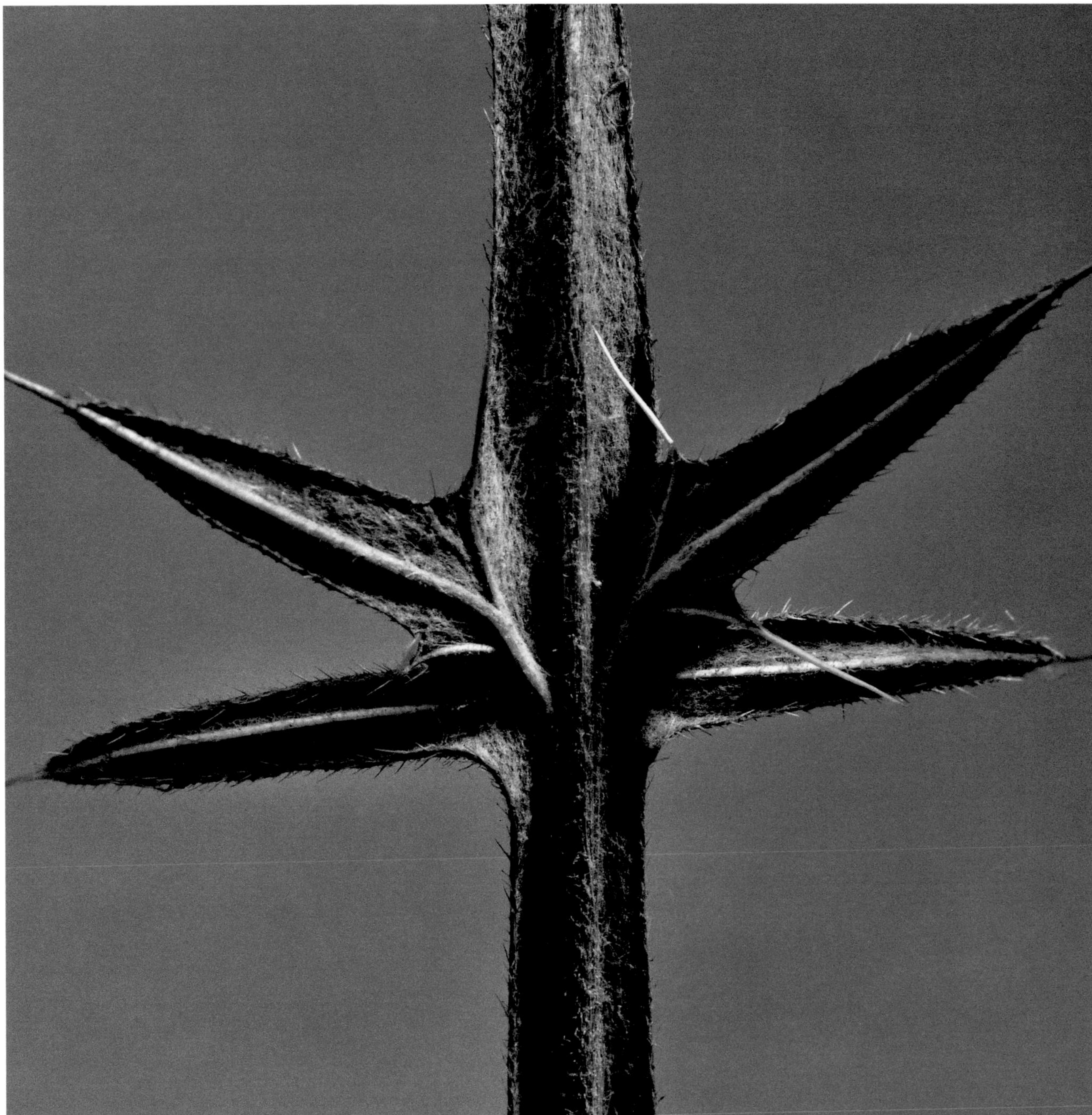

Cirsium sp.

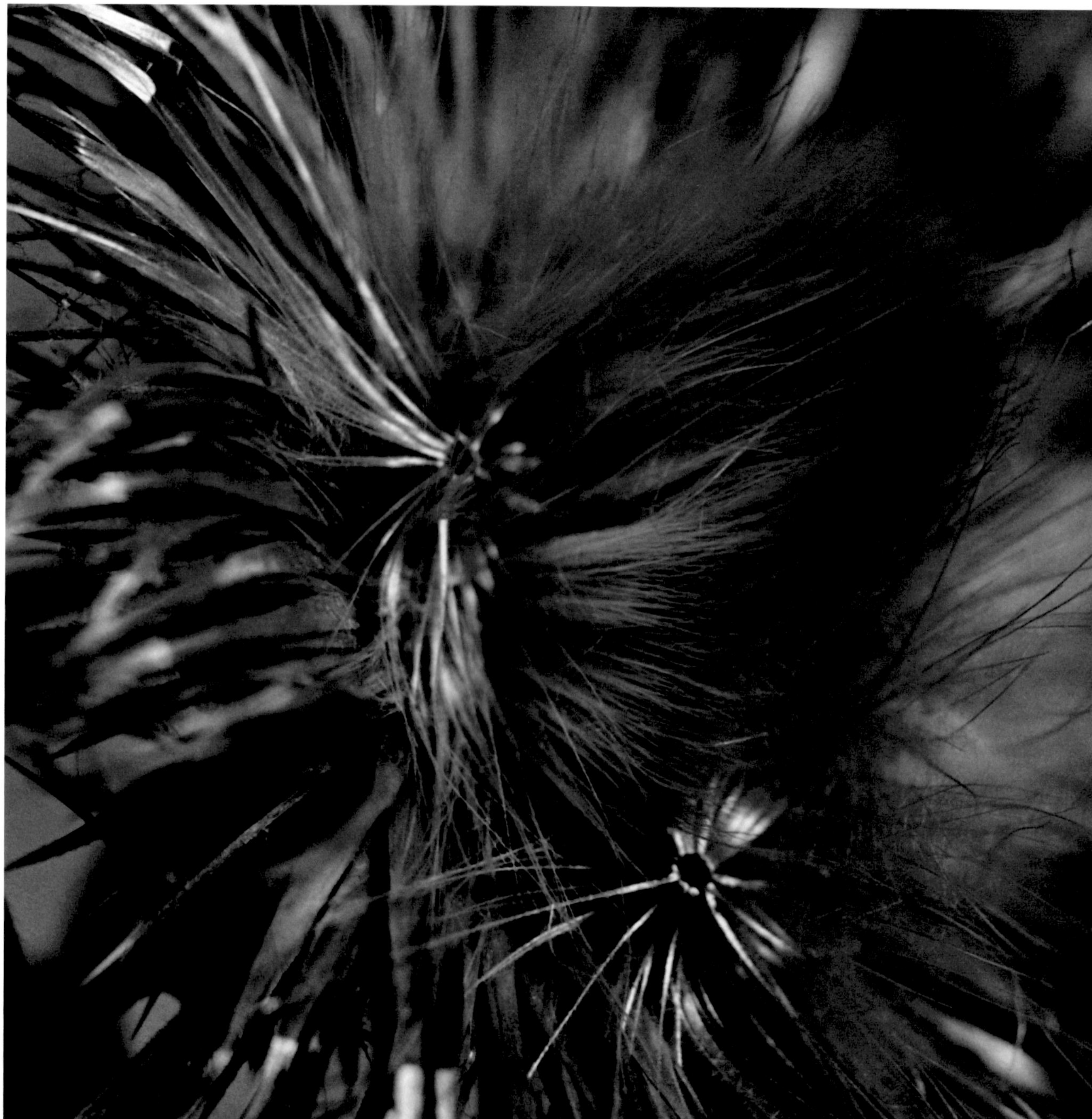

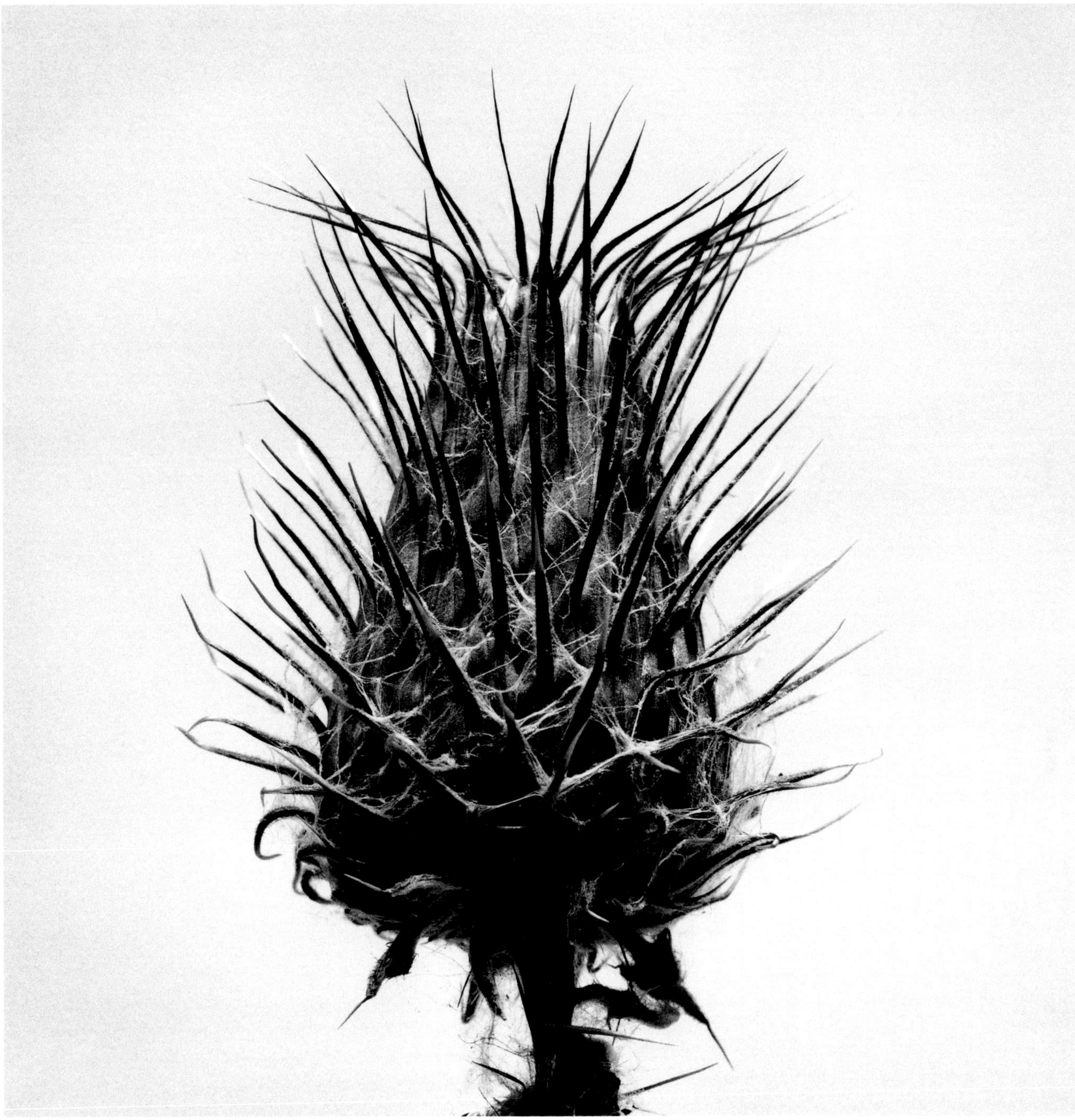

Cirsium sp.

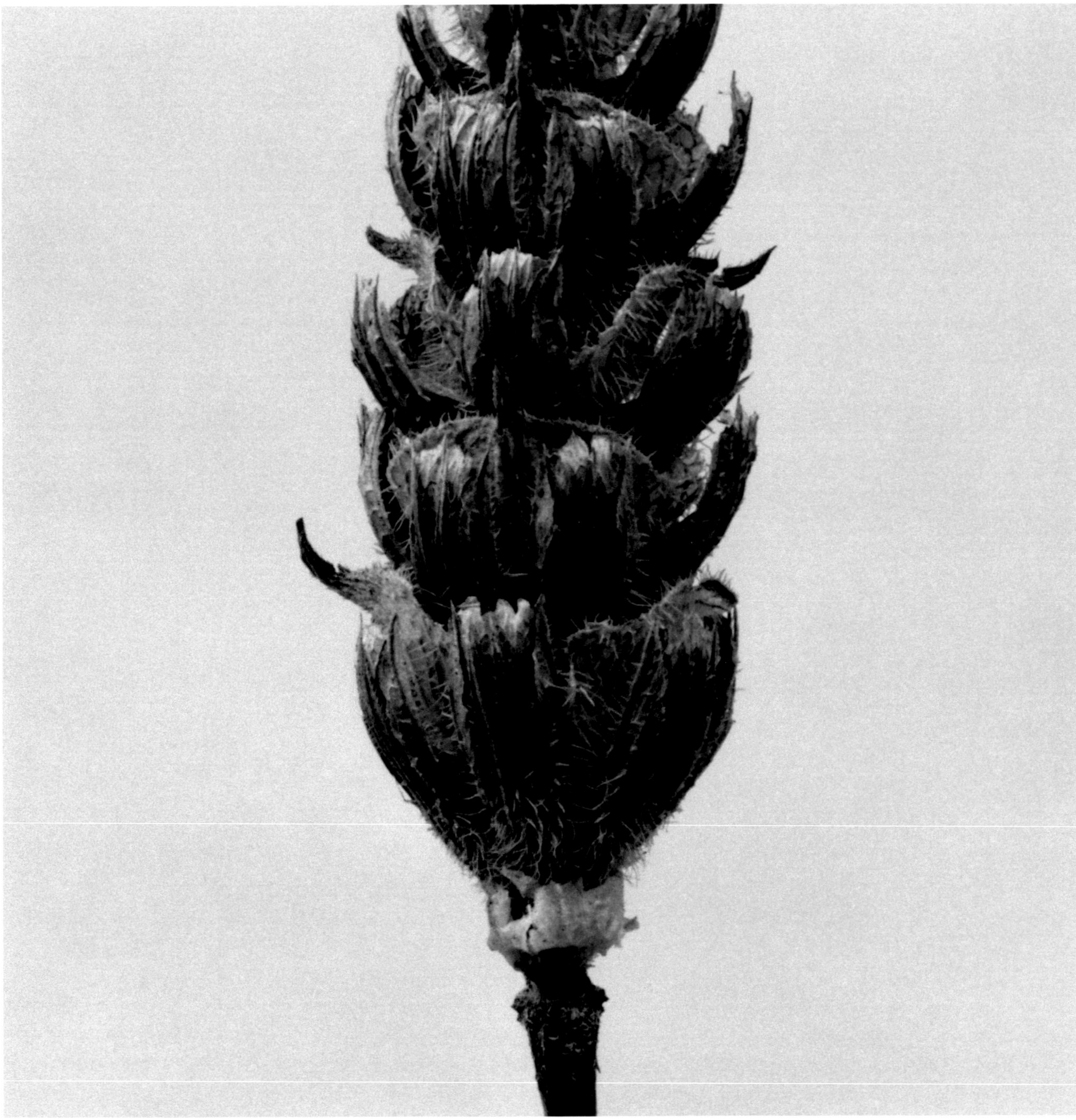

Prunella vulgaris

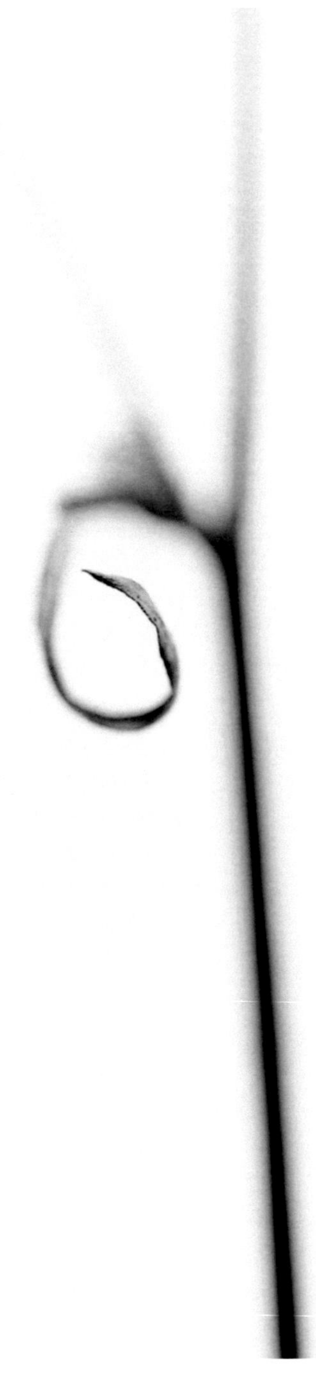

Prunella vulgaris

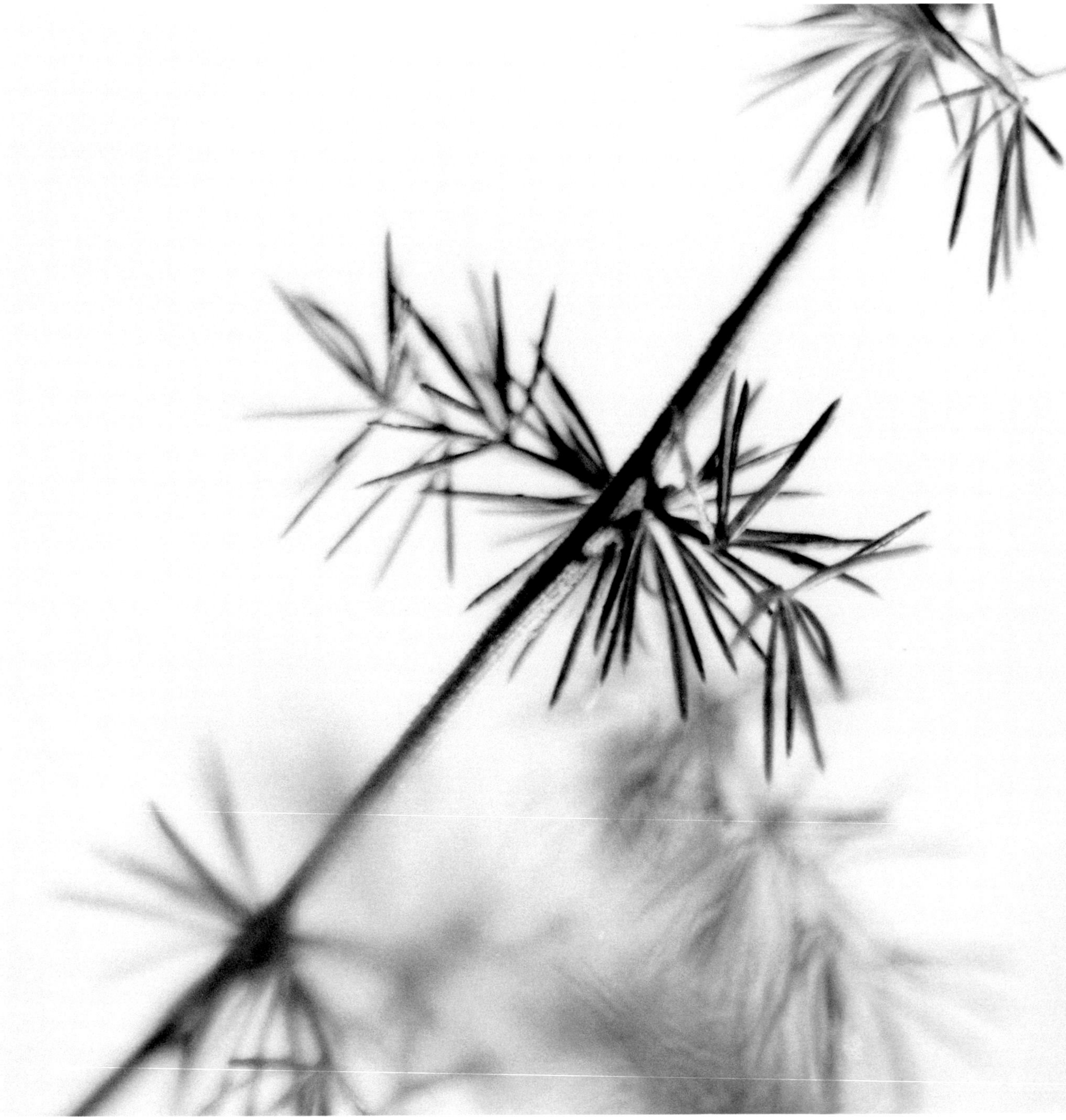

Asparagus officinalis

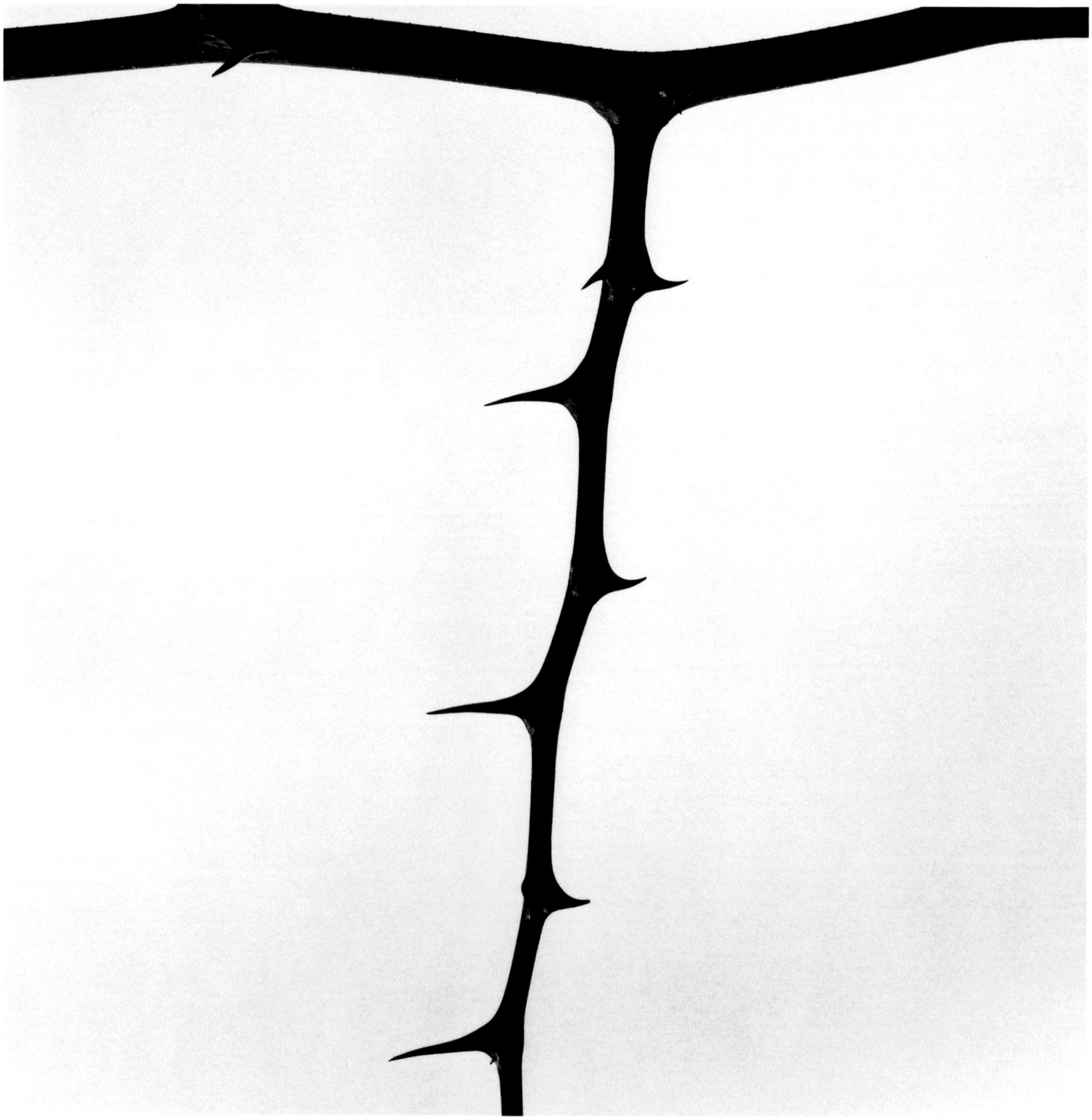

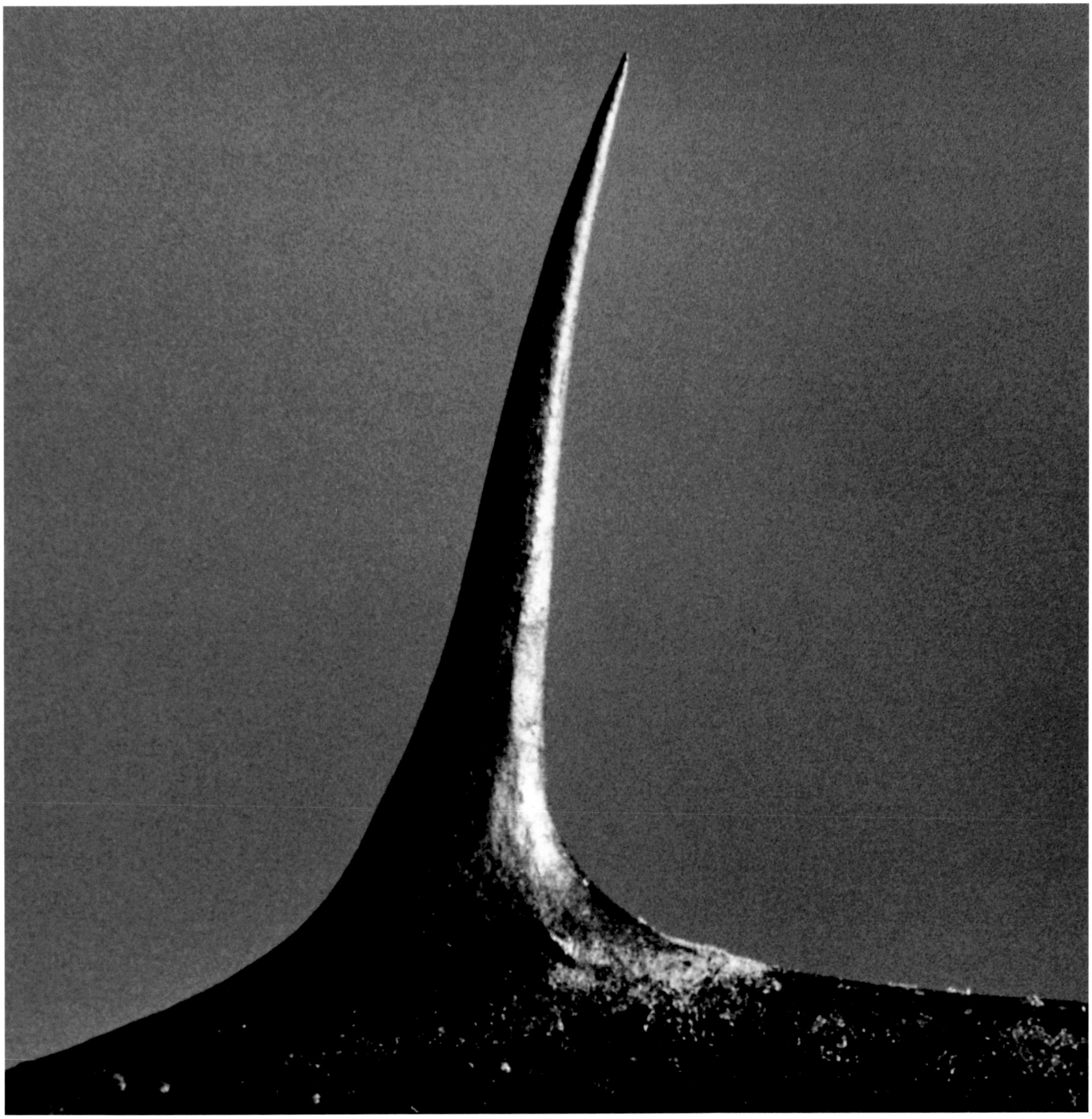

Paliurus spina-christi

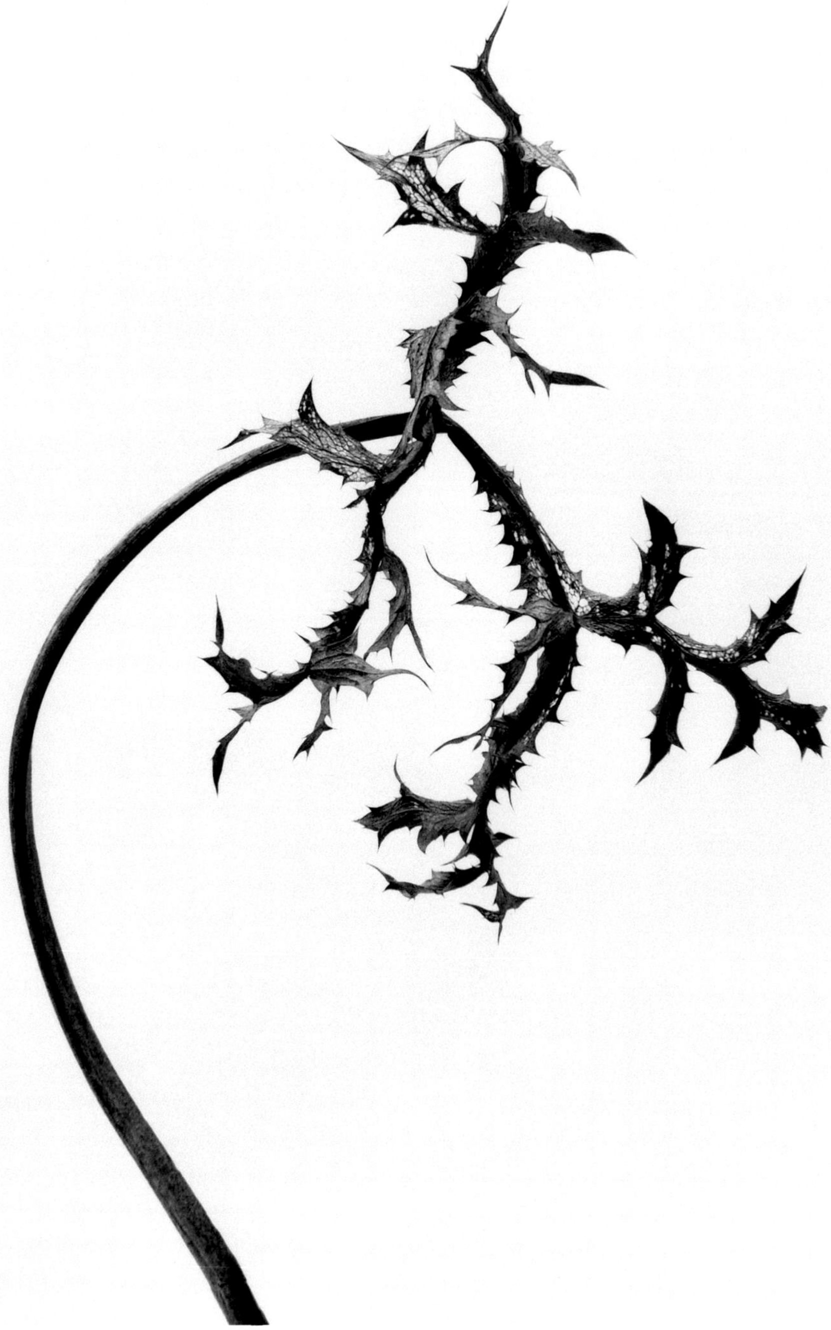

Eryngium campestre

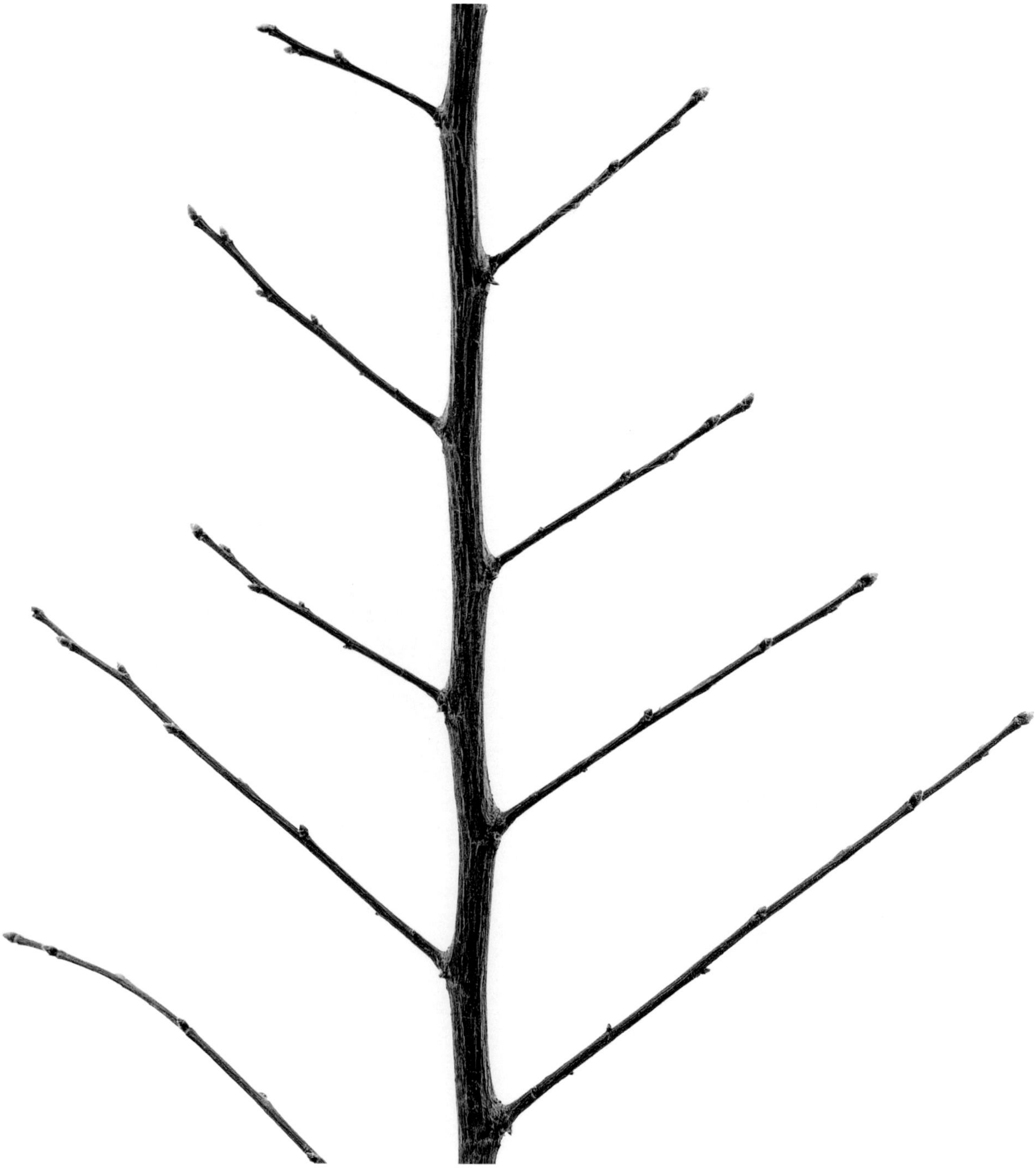

Ulmus minor

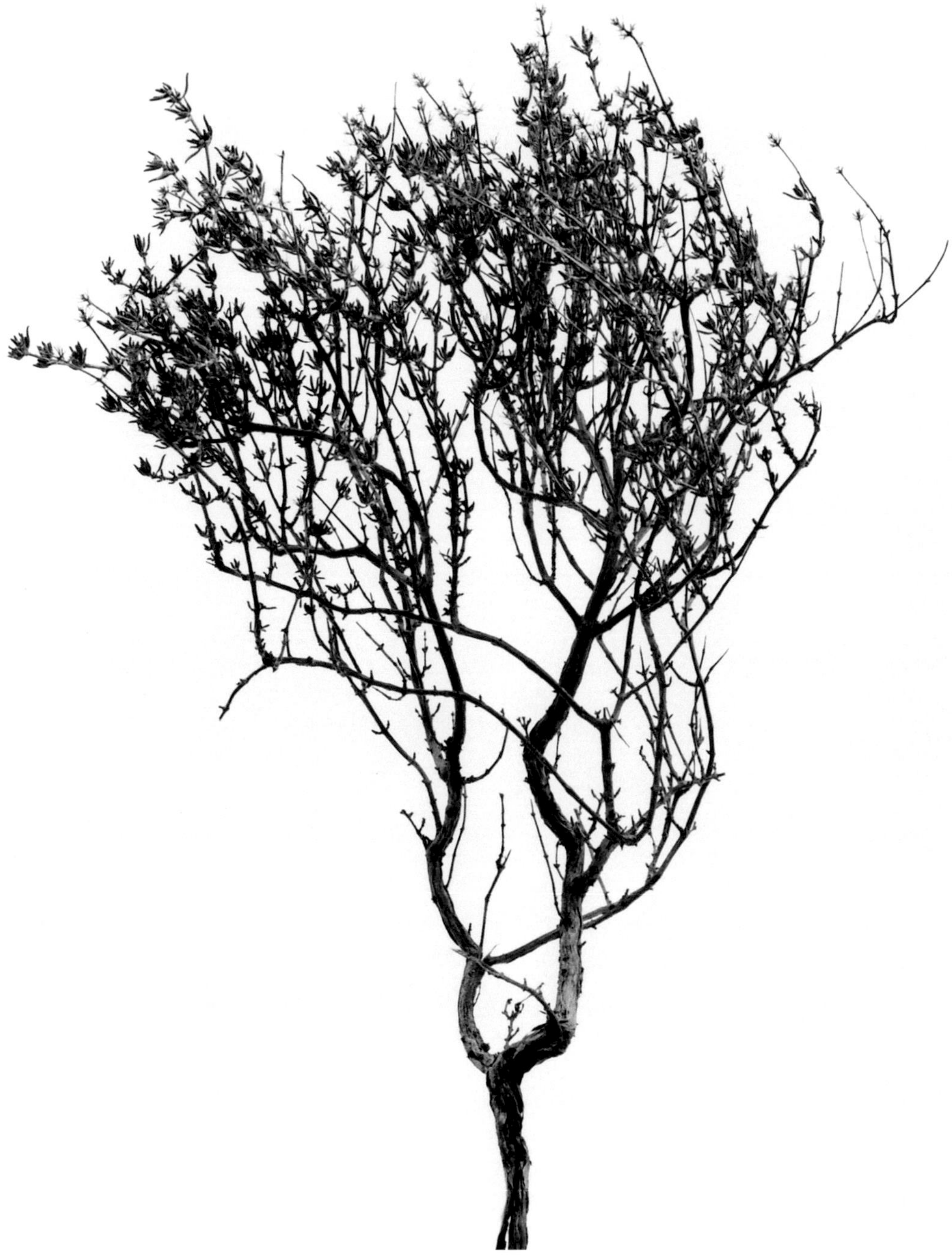

Thymus vulgaris

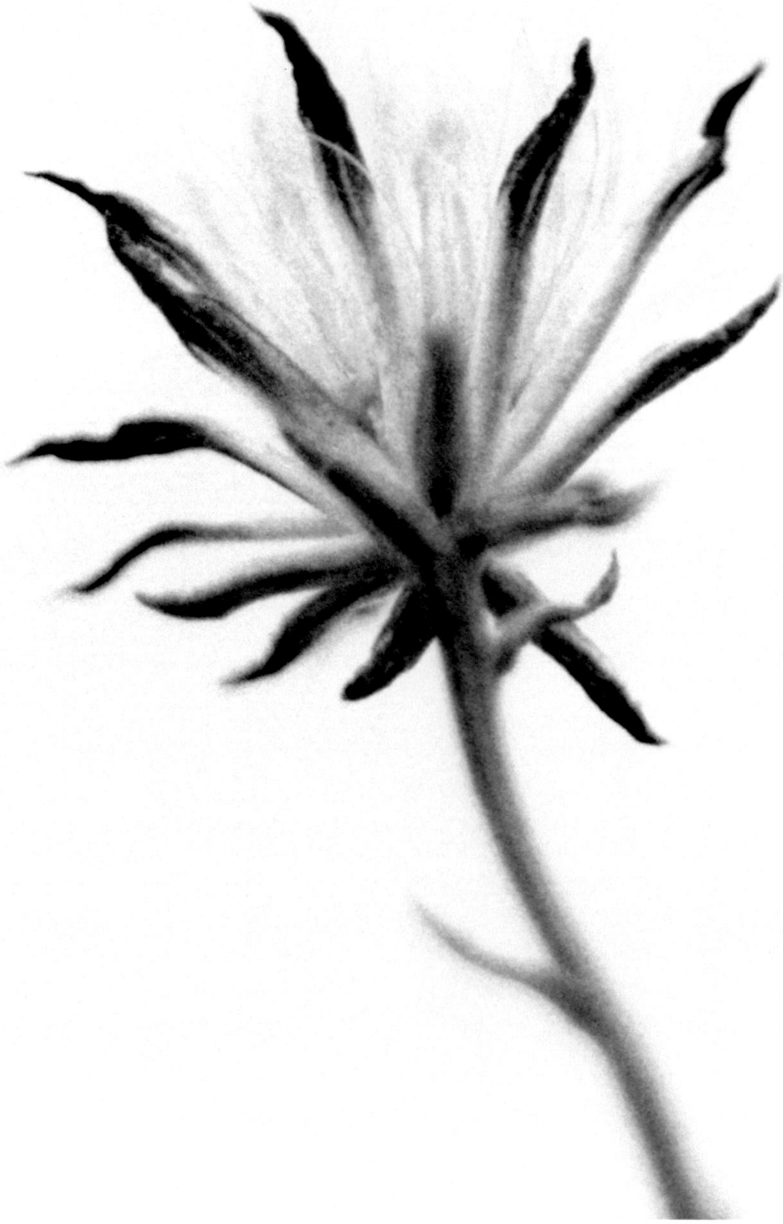

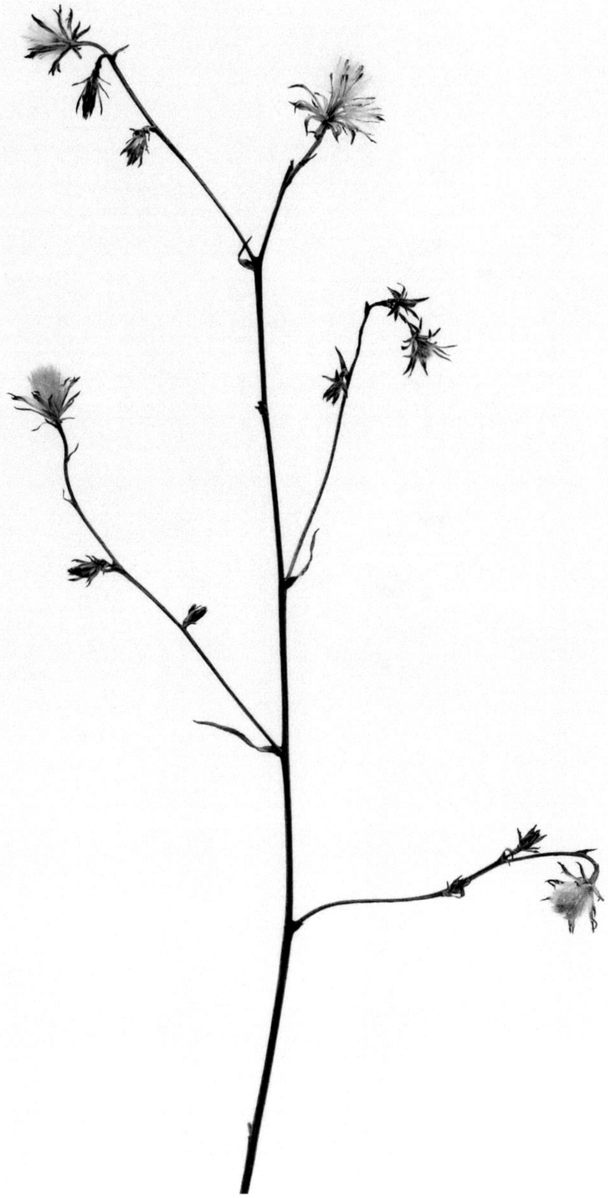

Aster pilosus

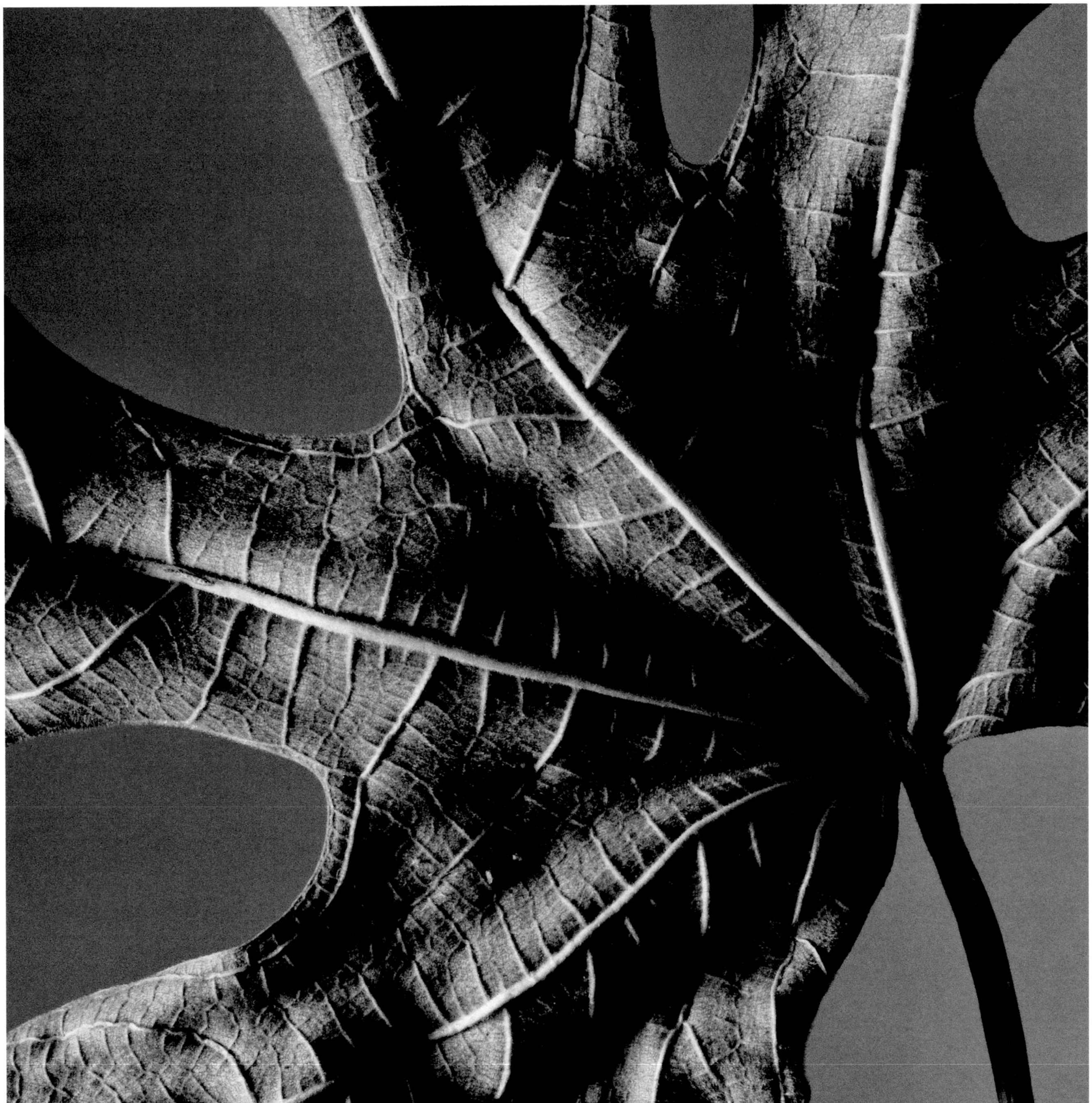

Ficus carica

Edition and Production: Turner
www.turnerlibros.com

Design: Silvia Rotllant

Colour Separation: Cromotex
Printing: Artes Gráficas Palermo, S.L.
Binding: Ramos, S.A.

Page 21: Masaoka Shiki, *Cien haikus*, Poesía Hiperión, 267, Madrid, 2004, p. 54. Translation by María Luisa Balseiro

ISBN: 978-84-7506-784-1
D.L.: M-4845-2007